Preparing Plant Tissues for Light Microscopic Study

A Compendium of Simple Techniques

Preparing Plant Tissues for Light Microscopic Study

A Compendium of Simple Techniques

RICHARD C. KEATING

MISSOURI BOTANICAL GARDEN PRESS

ISBN 978-1-930723-28-3
Library of Congress Control Number 2014953566

Monographs in Systematic Botany from the Missouri Botanical Garden
Volume 130
ISSN 0161-1542

Scientific Editor and Head: Victoria C. Hollowell
Managing Editor: Allison M. Brock
Associate Editor: Lisa J. Pepper
Associate Editor: Laura Slown
Press Coordinator: Pacia Anderson
Production Editor: Hilary Lord

P.O. Box 299
St. Louis, Missouri 63166-0299, U.S.A.
www.mbgpress.info

Front cover image:
Eugenia uniflora, cleared, with veins stained in iodine, mounted in calcium chloride. See "Hydroxide, Bleach, and Iodine" in chapter 3. Specimen furnished by Dr. Neil Snow.

The photos listed below show transverse hand-sections, stems or petioles, stained in dye solutions and mounted in aqueous calcium chloride. See chapter 1: "Calcium Chloride Solution as a Mountant for Stained Hand-Sections."

Color image endpage insert:
A. *Epipremnum* in safranin. —B. *Epipremnum* in Schiff's. —C. *Philodendron* in hematoxylin. —D. *Epipremnum* in hematoxylin. —E. *Epipremnum* in cresyl violet acetate (CVA) with cork. —F. *Epipremnum* in thionin. —G. *Epipremnum* in cresyl blue. —H. *Epipremnum* in CVA.

Back cover image:
Acalypha hispida t.s. stem

Contents

Preface

Following my early retirement from Southern Illinois University-Edwardsville I was offered laboratory space at the Carbondale campus of Southern Illinois University. For this I am indebted to Professor Lawrence Matten, then chairman of Plant Biology.

While at SIUC, in addition to conducting research in structural botany, I had the opportunity to teach my course in plant anatomy for a final time. Coincidentally, I had the good fortune to have learned some excellent hand-sectioning and slide-making techniques from the late Professor Joseph Varner of Washington University in St. Louis and from Professor John M. Herr, Jr. of the University of South Carolina.

Using those exciting tools, new to me, I completely revamped my methodology for teaching this subject. While the lecture portion was based on the usual textbook sequence, the laboratories were entirely different. Using easily taught new techniques, the students made all of their own slides from campus or greenhouse materials. It was by far the most successful course I ever taught. Frequently, students remained far beyond the end of scheduled periods, eagerly showing each other their beautiful preparations. Since their work was not based on conventional paraffin microtechnique, they quickly became competent and could concentrate on the anatomy they were supposed to be trying to understand.

With this experience I bore down on ways to accelerate productivity in data-gathering using simpler approaches that involved the use of fewer tools and less space. Of course there are lines of inquiry, such as developmental studies, where the making of serial sections requires a well-equipped microtechnique "wet lab." However, to my great satisfaction, I found that much useful anatomical data can be harvested in surroundings of limited equipment and space. In more recent years I continued this work at the Missouri Botanical Garden and I am grateful for the generosity of its director, Dr. Peter Raven.

I am also indebted to many others as this project took form. First, I will never forget the inspiration I gained from my mentor, Professor Thomas K. Wilson, whose love of method was as great as mine became. Here at Missouri Botanical Garden Press, Victoria Hollowell, Head, was encouraging and helpful throughout the writing and editing process. Members of her staff, Tammy Charron and Lisa Pepper, were helpful in many ways. Fred Keusenkothen and Randy Smith rendered the illustrations into publishable form. Finally, I am indebted to two extraordinary editors: Ellen Kunkelmann, technical editor, and Hilary Lord, production editor, who moved the manuscript with good-humored efficiency through its final metamorphosis.

The present compendium of nearly 200 procedures and ideas is organized around my decades-old literature collection, and I have tried the majority of them. Also, I thank friends and colleagues who offered favorite techniques. I do note that this collection does not pretend to constitute an exhaustive survey of the literature. Not all techniques I reviewed made the cut. Procedures that survive here are capable of yielding very informative preparations. Techniques not included had either marginal efficacy or unnecessary complexity or were unnecessarily dangerous to your health.

I am convinced that many new or modified methods of tissue preparation can yet be developed and hope that this guide will inspire such experimentation.

Introduction

The best preparation technique is the simplest and least drastic that will yield the target data.

—Keating's corollary, Occam's razor

The purpose of this guide is to provide references and basic instructions for simple preparative techniques. The user I have in mind is a teacher, researcher, or student who has some background in the interpretation of plant structure, and who has a research question that could be informed by the collection of light microscopic data.

I imagine investigations by those with limited spacc or those who have no access to a well-equipped microtechnique laboratory. Also, where feasible, I emphasize the use of supplies that can be made simply or obtained from local businesses. It is not always necessary to rely on expensive scientific supply houses. I do assume the availability of a clinical- or research-grade light microscope (LM).

In a few cases, techniques relevant to the use of the scanning electron microscope (SEM) are mentioned, but this instrument is mostly outside the focus of this guide. Many of the techniques in this guide require some manipulative competence, but the requirements for tools and commodities are usually limited.

A number of texts and reviews provide excellent coverage of general microtechniques involving specimen embedment, microtomy, multiple staining, dehydration, and resin-mounting; for example, Johansen (1940), Gatenby and Beams (1950), Gray (1954, 1958), Sass (1958), Purvis et al. (1966), Feder and O'Brien (1968), Sumner and Sumner (1969), Willey (1971), Berlyn and Miksche (1976), O'Brien and McCully (1981),

Ruzin (1999), and Sandoval and Rojas Leal (2005). Chamberlain (1932) is also a good source for earlier methodology.

Techniques involving embedding and serial sectioning remain essential for developmental investigations, or any studies wherein sections of known thickness are required. Except for tangential mention, this guide is not about that kind of work; that niche is already filled.

In this era of accelerating and ever-more powerful technology, a compendium of simple, often older techniques might seem like a retrograde notion. But I hope to demonstrate that such approaches remain expeditious for gathering many kinds of microscopic structural data from plants. Many such techniques are nearly forgotten or obscurely published. Many have been superseded over the years by methods requiring major investments in laboratory space, equipment, or expensive commodities. Users of this guide will also find many useful procedures in Peterson et al. (2008).

Absent experience in manipulating plant tissues, it saves much time if you can arrange to watch an experienced worker. Otherwise, plunge in and practice with your local plants before moving on to rarer research material. The techniques included here allow the harvesting of observations from hand-sections, replicas, clearings, and wet mounts. Useful observations will include the presence or absence of cell types, their form, abundance, distribution, or counts per area, observations where serial sections are not required.

Many wet mount techniques, especially those outlined in chapter 1, are easily used in the field. My portable laboratory is a small tackle box, which carries slides, blades, dissection tools, and dropper bottles. With this kit and a small microscope, my curiosity has led me to make preparations on a picnic table in a campground or in a cabin at remote locations.

Procedure modifications. When manipulating plant materials you will often need to experiment. Solution concentrations, time periods, and temperatures listed are what have worked for some authors, for some species. For many plant groups, optimal methodology is unknown. On occasion your specimens will pose some different problem for hand-sectioning, clearing, or whole mount preparation. Therefore the various treatment times and fluid concentrations should be regarded as starting points. For instance, when making macerations, some specimens are quite hard and resistant to treatment while others have tissues that practically dissolve if left for the recommended period. You need to monitor the process for each new type of material. Be sure to keep good notes using a research notebook or your own data forms.

Editorial comments and notes, given in the first person, are my opinions based on my experiences. A pet peeve: If you publish your work, give the reader the key details. Many authors are too vague in this regard, or they cite sources that are also vague.

Citation practice. In spite of my attempt to give the techniques of others in sufficient detail, please remember to cite the original author. When possible, the original reference should be consulted as it may contain a more detailed review of variations and technique subtleties.

Safety disclaimer. The user of this guide is responsible for understanding the hazards of using chemicals and any techniques listed within. Occupational Safety and Health (OSHA) Guidelines are available, and there are Material Safety and Data Sheets (MSDS) for each chemical from each manufacturer. They are available on the Internet, and you should consider it your responsibility to understand their contents. Similarly, sectioning with razor blades can be hazardous. As with power tools, be fully awake and focused when using them.

Be aware that a number of chemicals, such as ethyl alcohol or chloral hydrate, require special institutional licenses before placing an order. Also, when a published procedure calls for the use of toluene or xylene, I usually cite it as published. However, I have found that D-limonene (in such products as Histo-Clear, Clearene, or Hemo-De) is a much safer alternative and can be substituted with no noticeable downside for those ring solvents that are unsafe to breathe. Safety sheets note that limonene can become a skin allergen, thereby reinforcing the need for habitual care in handling all chemicals.

Finally, Bushnell (2013) has called attention to the subtle dangers of breathing various common solvents, such as ethanol and toluene, including their physiological effects on attention, such as when driving after exposure.

CHAPTER 1

Dissection, Hand-Sectioning, Fixing, and Tissue Printing

The following hand-sectioning, staining, and wet mounting techniques are useful for softer materials. Fixation produces better slides in some cases, although fresh specimens may also yield excellent preparations. Woody and other harder materials are discussed in chapter 2, but tissue printing, described here, will often work well with either kind of material.

Handmade Cross Sections of Leaves

(J. Varner, personal communication). Using the following technique, very informative, uniformly thin cross sections can be made of leaves or other flat plant materials.

Procedure

1. Supplies: (a) A plastic cutting surface. The best is a number 4 LDPE plastic food container lid, as it will not dull razor blades; (b) Double-edged razor blades designed for shaving. These are the sharpest and will be used for the cutting; (c) A single-edged razor blade, the back of which will be used as a straight edge. Leave the protective cardboard guard intact; (d) Sharp-pointed dissection forceps; and (e) Watch glass, depression slide, or spot plate with 15% ethanol.
2. Take a strip of leaf, about 1 cm wide, and cover up half of it with the back of a single-edged blade. The rounded edge of the metal blade backing will be your straight edge. Hold it down firmly so that it will not move during cutting. With the double-edged blade held vertically, cut off a piece of the leaf. (See Figure 1.)

3. Without moving the specimen or the straight edge, make another cut holding the double-edged blade at a slightly oblique angle so that it undercuts the previous cut. Since the convex-curved back of the single-edged blade is about 0.5 mm above the leaf surface, this second cut will produce a thin cross-sectioned strip of leaf. Use the forceps to pick up the fresh thin section and place it in the spot plate, depression slide, or watch glass containing some drops of water or 15% ethanol. Dyes can be applied and rinsed at this stage.
4. When this cut piece is placed on a slide with water and a cover glass, it will appear remarkably flat and uniform in thickness. Because leaves already have pigmented plastids, no additional stain is usually necessary. Stains are useful if one needs to differentiate the ground tissue around midveins, or for detection of lignified tissues, collenchyma, or other materials.

Softening Hard Leaves Before Sectioning

Zuloaga and Morrone (1996) made good hand-sections of grass leaves after treatment with hydrofluoric acid as follows: Leaf pieces, dry, fresh, or preserved in formaldehyde alcohol acetic acid (FAA), are soaked in Contrad 70 (Decon Labs, Inc., King of Prussia, PA) for 24–48 hours at ambient temperature. Pieces are transferred to 5% HF for 24 hours. Presumably they are rinsed.

Cross sections are made and stained in 1% methylene blue and 1% safranin in 80% ethanol. Alternatively, safranin-alcian blue can be used.

Paradermal Preparations of Leaves

(Keating). It is possible to make paradermal sections that are quite adequate for determining the nature of epidermal cells, stomatal type and distribution, crystal type and distribution, and other characters.

Procedure

1. Supplies: (a) Two pieces of Styrofoam about 5 mm thick and 1 cm^2; (b) Double-edged shaving blade that has been snapped with pliers

at the ends of the slot to make two, single-edged pieces; and (c) Four leaf sample pieces, cut from the same leaf, about 1 cm^2.

2. Wet the leaf surfaces. Make a sandwich with the four leaf pieces stacked between the Styrofoam pieces, like a four-layered hamburger.
3. Holding the single-edged blade piece by one end, saw through the stack, paradermally, at a slight oblique angle so as to get obliquely cut paradermal sections. If necessary, do this several times, remaking the stack between cuts, to get enough sections. It may be helpful to hold the stack against an LDPE cutting surface when sawing through the stack.
4. The resulting cuttings will be irregularly sliced shards when mounted under a cover glass. The obliquely cut edges will yield information on the paradermal leaf structure. Best results are obtained when mounted in calcium chloride (see following section).

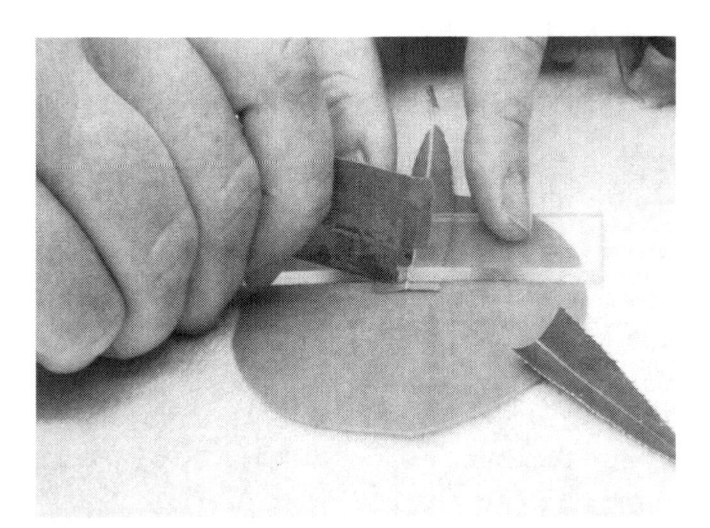

FIGURE 1. Cutting sections against number 4 LDPE plastic. For a straight edge, one can use the back side of a double-edged razor blade. Alternatively, one can make a straight edge using a piece of ⅛-inch acrylic plastic. The edge is asymmetrically ground (sanded) to a 45-degree angle on both sides. Most of the slope is on the top side. For the bottom side, grind a bevel to about ½ mm deep. To make a semicircular groove to protect the midrib, clamp together two pieces of plastic and drill a hole at the boundary between the two pieces.

In various *Anatomy of Monocotyledons* volumes, Metcalfe, Tomlinson, and Cutler used a scraping technique to look at epidermal features. This description is summarized from Cutler (1969).

To examine a leaf surface, place the leaf onto the surface of a glazed ceramic tile; surface to be examined face down. Flood the piece with household bleach (i.e., ca. 6% sodium hypochlorite). If previously dried, the sample has been restored before this process. To keep one's hands out of the bleach, hold down the leaf piece with a bottle cork. Scrape away tissue with a razor blade until the lowest epidermal layer has been reached. Transfer the leaf piece to a small dish and continue soaking in bleach for 10 minutes. Rinse with water and remove loose pieces of debris with a camel-hair brush. Leaf piece can be stained and mounted as required.

NOTE: This technique should work for hard-surfaced leaves, as found in many monocot species, as well as for coriaceous dicot specimens.

Calcium Chloride Solution as a Mountant for Stained Hand-Sections

The following technique contains elements of sectioning, stain use, mountants, and microscopy. It is placed here since it works best as an integrated description.

Microtomed serial sections will continue to be necessary for study of vascular patterns and development, apical meristems, and buds of various types, where growth processes are being investigated and where known section thickness is required. However, where screening for histological features is the issue, hand-sections and well-made wet mounts of small whole organs remain informative for many data-gathering purposes.

(Herr, 1992; Keating, 1996, 2000, and unpublished; Ogburn & Edwards, 2009). Except for the teaching laboratory, the use of wet mounts in the study of plant anatomy has never been as common as the making of microtomed sections of embedded materials. Certainly, permanent resin-mounted slides comprise a growing reference collection over the years, as they store archivally and ship well.

Herr proposed the resurrection of the use of calcium chloride solution as a cover glass mountant (1992). He reviewed its early use, later abandonment, and replacement by glycerol for wet mounts before the introduction of aniline dyes in the late 19th century.

As with all alcohol-based mountants, tissues mounted in glycerol make satisfactory microscopic preparations, but they are suboptimal. Many dyes tend to leach from the specimen into the solution, or their metachromatic or differentiation properties are destroyed. Until Herr (1992) separately tested toluidine blue O (TBO) and phloroglucinol in calcium chloride, no other dyes had been so evaluated.

Although several dyes such as TBO are somewhat metachromatic in water mounts, they are temporary. In water mounts of plant tissue sections, TBO or phloroglucinol will differentiate lignified tissues well.

Procedure

1. Overview: Make a tissue section (see "Handmade Cross Sections of Leaves," above). Immerse it in dilute ethanol or fixative, rinse, transfer to dye, rinse in dilute ethanol, then mount in calcium chloride solution.

   NOTE: Especially before mounting wood sections, let the sections sit in a small dish or watch glass of dilute ethanol for about 15 minutes. They can then be mounted in calcium chloride solution with few or no bubbles trapped beneath the cover glass.

2. After sectioning, place a few drops of stain on a depression slide, spot plate, or watch glass. For best results, use cresyl violet acetate (CVA), but see discussion below. With forceps, place a section in the stain and wait 2–4 minutes. Transfer the section to several drops of 15% ethanol and leave for several minutes. This leaches away unbound dye molecules.
3. Transfer the section to a slide; it should look somewhat opaque and unpromising at this point. Add several drops of calcium chloride solution and add a cover glass. The process can be monitored on the stage of a dissection microscope. Metachromasy will show up immediately, but it may take 30 minutes before the mountant becomes glass clear.

In fact the preparation looks its best after several hours when the high refractive index (RI) salt solution has penetrated all interstitial space.

4. For light microscopic study (LM), set up the microscope lamp and field diaphragm for Köhler illumination (see chapter 8), but with the substage diaphragm wide open. The high RI of the mountant ensures that you can focus on the cut surface of a somewhat thick section without visual interference from out-of-focus tissues below.

Tests of the Calcium Chloride System

Sections. To make comparisons, I made about 100 thin hand-sections of fresh stems of *Epipremnum* and *Bougainvillea* from greenhouse-grown plants, placed them in FPA_{50}, and stored them in 70% ethanol. In some cases I stained and mounted fresh sections directly, but preparation clarity is improved if the sections are fixed briefly or first stored in dilute ethanol. Specimens restored from dried material give the same stain reaction as long as you avoid using chloral hydrate, which destroys stain differentiation in any subsequent procedure. There was little difference in stain reactions between fixed and restored material, only that fresh or fixed materials will show better cellular contents.

Stain solutions. As part of the investigation, I decided to see if measured proportions were really necessary. They aren't, as explained below. A microspatula of each dye was placed in a 4-dram vial and 15% ethanol added. The stains that worked were readily soluble in aqueous solutions, but a small percentage of ethanol reduces surface tension and improves penetration of dyes into tissues. When held up to the light, the stain solution in a vial should appear barely translucent. This works out to be between 0.01% and 0.1% solution when compared with measured quantities. The optimal stain time is determined by experience, which varies somewhat among disparate plant materials. If sections take up stain within seconds and become uselessly opaque, dilute the stain solution. Stains penetrate inward from the cut surfaces.

Mountant. Calcium chloride should be made up by adding and stirring the salt gradually into a beaker of water. The cheapest, technical-grade salt is all that is needed. Anhydrous versions of the salt often do not dissolve well. Also, they are highly exothermic when water is added; adding water quickly could crack the vessel.

A solution between 20% and 40% salt works well. The exact concentration does not matter because, over several days, excess water evaporates from the edge of the cover glass until the concentration of this hygroscopic salt solution equilibrates with atmospheric humidity. After 24–48 hours, as necessary, use a transfer pipette to add some of the mountant solution to the edge of the cover glass to make up for evaporation.

Hardening the mount. I have kept numerous slides, stored flat, for years without deterioration, but they continue to need the careful handling of any wet mount. For more resistance to tilting, some researchers resort to the time-consuming method of cover-glass ringing.

In experiments with hardening the mount itself, I had best results when adding polyethylene glycol to the calcium chloride solution. It is miscible in such aqueous solutions and its RI is appropriately high. It is sold over the counter as MiraLAX (Merck, Sharp & Dohme, Kenilworth, NJ), a treatment for irregularity. One can experiment by adding a drop of polyethylene glycol solution to a calcium chloride mount. Start with a 10% solution. Too much will cause the mount to lose its transparency.

Calcium chloride properties. $CaCl_2$ is a very hygroscopic white powder that can be obtained in various hydration states. It takes ca. 74.5 g/100 ml H_2O to make a saturated solution. Water's dielectric constant is high (80), which separates the Ca++ and Cl– ions to form hydrated ions, a strongly exothermic reaction when mixed with water. The salt is heavy (specific gravity 2.15) and also liberally soluble in ethanol or acetone.

Its most interesting property when used as a slide mountant is that its RI matches that of crown glass: 1.52 (Haynes, 2010). This raises the preparation's numerical aperture and clarity (see www.microscopyu.com). Specimens mounted in calcium chloride are remarkably clear, with stained tissue appearing as translucent as stained glass.

As long as calcium chloride wet mounts are stored flat, such preparations will last for years or indefinitely without drying out, and the stains won't lose color. Oddly, for such a polar solution, the salt does not cause cell or tissue shrinkage or collapse of cell contents of fresh, fixed, or restored specimens. It was this combination of properties that caused me to investigate a variety of dyes to see if hand-sectioned specimen preparations could be improved.

Stains. Beginning with the quinone-imine class of metachromatic stains (CVA, thionin, methylene blue, and TBO), I screened a large collection of stains common to plant or animal histology laboratories. Several of the stains discussed below turned out to have outstanding potential for permanence, brilliance and clarity, and sharp differentiation when mounted in calcium chloride solution (see Plant Tissue Dye Responses Table, page 130). Any tendency toward metachromasy is accentuated. Most of these stains have never been mentioned in combination with calcium chloride in the literature of botanical microtechnique (Keating, 1996).

NOTE: Dye nomenclature can seem confusing. I have used the preferred spellings and suffixes given by Lillie et al. (1969) that were approved by the *Commission on Standardization of Biological Stains.* Seemingly minor molecular variations can lead to major differences in absorption spectra, hence the specific spellings and suffixes. Historically, however, commercial dyes have been shown to vary in purity. Also, the terms *dye* and *stain* are often used interchangeably.

Test Results

The stains listed in the Plant Tissue Dye Responses Table at the end of this manual represent, in order of merit, those that have the best potential for providing informative preparations. That is, they provide sharp differentiation and clarity. The number of plus marks (+) under metachromasy and differentiation are a subjective ranking of the strength of the stain reaction. Described below are those ranked highly useful.

Cresyl violet acetate (CVA). Little known among botanists, this dye has been used by animal histologists as a stain for nerve cells (Conn et al., 1960). Probably for the first time, Dizeo de Strittmatter (1980) reported on the use of this dye applied to plant materials. However, her specimens were mounted in glycerin/gelatin and the preparations did not obtain the degree of metachromasy that I noted later when CVA was used with the calcium chloride-based mountant.

In testing dozens of stains, CVA produced the most outstanding results of any so far investigated. Its preparations are of great clarity, differentiation, and metachromasy. Its polychromic range of color reaction frequencies is wider than with any other stain (see Table).

After rinsing in an ethanolic solution, CVA-stained sections will appear uniformly violet–not bad, but with no differentiation and lacking interest. If you add the calcium chloride to the section while observing through a dissection microscope, the transformation is dramatic. Within seconds, parenchyma walls turn tan, while lignified walls become bright blue (see also Ogburn & Edwards, 2009).

It is important to rinse all excess dye from sections using dilute ethanol before adding mountant. Otherwise, unbound remaining dye molecules will form polymer threads or clusters, artifacts that could spoil the specimen's appearance.

The other dyes of the quinone-imine class are useful but results are not as striking as with CVA. An interesting feature of lignified walls is that some specimens having both xylem and extraxylary fibers will show different color reactions in the two tissues, such as blue in xylem and blue-violet in fibers. Herr (1992) also illustrated this effect (without comment), where phloroglucinol looks dark or lighter red in different lignified tissues. Rogers et al. (2005) note a similar color response and consider this to be a qualitative assessment of the proportions of syringyl (S) and guaiacyl (G) monolignols. In angiosperms these subunits, in various proportions called the S:G ratio, build the lignin polymer (see Whetten & Sederoff, 1995; Eckardt, 2002). This stain/mountant system described here is probably detecting variation in these lignin moieties in xylem and sclerenchyma walls, and this needs further investigation.

Iodine-potassium iodide (I_2KI). While iodine and potassium iodide are not dyes, when used with calcium chloride this element/salt combination does more than detect starch. The lignified tissues become yellow/orange in striking contrast to the magenta of cellulosic walls, a variation of Artschwager's (1921) zinc-chlor-iodide test for cellulose.

The effect develops over several days in an excess of I_2KI. Unlike aniline dyes, I_2KI is fully soluble in calcium chloride. In fact, I_2KI can be added to a supply of the mountant, thereby combining staining and mounting. Alternatively, after removing the section from I_2KI, it should be mounted directly in calcium chloride with no concern about excess carryover into the mountant. The aqueous I_2KI solution: Iodine: 0.1 %; potassium iodide: 1.0%.

Other stains. The stains listed in Plant Tissues Dye Responses Table show at least some differentiation except for phloxine. Those listed in footnote 2 of Plant Tissue Dye Responses Table had no merit compared with unstained material. Other stains not listed produce muddy preparations, probably due to solubility issues or other chemical incompatibilities in this system.

Hematoxylin is a natural dye extracted from wood of the leguminous genus *Haematoxylum*, rather than from the usual benzene series-derived coal tar dyes. It produces strongly polychromic preparations that are unfortunately temporary. For up to 2 or 3 weeks, the red or orange lignified walls differentiate strikingly and photogenically against gray cellulose, but after that time the color is lost.

Crystals. For at least 1 week, calcium oxalate crystals appear normal and are easily detected, as usual, with the help of polarizing filters. But, within weeks, they will begin to erode in the calcium chloride mountant. If oxalate crystal morphology is important longer term, some sections should also be mounted in glycerin or glycerin jelly.

Preparation Longevity

My initial slides, about 20 years old, remain informative. They are stored in slide boxes so that the slides remain flat. There appear to be no limits on their value over many years. In case of a tilted or damaged slide, the section can be easily remounted with no loss of value.

Summary

Assuming some skill in making hand-sections of plant tissues, this dye and mounting medium combination will result in potentially permanent mounts for the study of plant tissues. Plant sections stained with CVA, and then mounted in calcium chloride solution, produce striking differentiation and polychromic effects. Originally suggested by Herr (1992) for use with TBO or phloroglucinol, this investigation evaluates numerous other dyes, many of which produce excellent results.

Tissue Dissection

(Romberger, 1966). Microdissection tools are especially useful when excising and manipulating apical meristems and other fine structures. They can sometimes be made more cheaply than when starting with expensive surgical instruments. Also, razor blade chips, insect pins, and fine sewing needles can be cemented to small dowel handles, and all can be used effectively.

Under a dissecting scope, Romberger had taken fine drills and hypodermic needles and ground them with dental drills to very fine diameters. Microforks can be ground by working opposing sides of hypodermic needles. They should be polished with fine rouge paper, or on end-grain boxwood, for microsurgical work. I have also used the grinders or sanding disks on Dremel tools (Robert Bosch Tool Corporation, Mount Prospect, IL). See Romberger's paper for illustrations and more details, or just experiment.

Dissection Tools

A note about leaf surface problems, especially those that must be cleared because they defy replica techniques. Many of my arctic *Salix* leaves are wooly, having deeply tangled trichomes forming a barrier up to 500 μm deep over the abaxial epidermis. Tape will not pull them off. But, using a light touch, they can be shaved or scraped from a dry leaf before clearing. Tools may be made as follows.

Making a scraper. Walter Sundberg (personal communication) has demonstrated, in class handouts, how to make dissection tools that are especially useful. Select a single-edged razor blade, the kind that comes with a cardboard edge protector. Leave the cardboard protector in place and, using a scalpel or sharp screwdriver, pry off the metal back of the blade. With the safety cardboard still in place, mark it diagonally with a pen in about 5 or 6 places. Using square-nosed pliers, break the blade into several fragments.

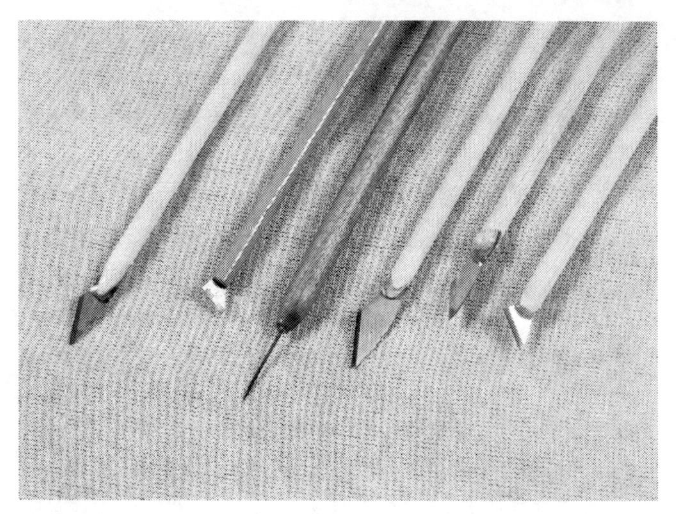

FIGURE 2. Cutting tools made using fragments of double-edged razor blades epoxied to wooden handles. Fine-pointed dissection needles can be made by cementing the distal half of insect pins into the ends of $^{1}/_{8}$-inch dowels. Drill end holes using a pin drill.

These blade pieces can be mounted in the ends of wooden dowels, $^{1}/_{8}$- or $^{3}/_{16}$-inches in diameter and 4 or 5 inches long. With a jeweler's saw blade, cut a short slot in one end of each dowel and cement the blade piece, at an angle, using epoxy. The resulting tool functions very well for a variety of dissection or scraping tasks. Sundberg placed these tools in disposable pipette bags so that the tools can be autoclaved any number of times. He used them for sterile work, such as removing spores from basidiocarps for cultural work.

Flower Dissections

(J. Pruski, personal communication, from techniques developed for making slides of *Asteraceae* florets by José Cuatrecasas and Angel Cabrera). This technique involves boiling florets, dissection, bleaching, staining, and mounting in glycerin. It is easy to prepare several taxa simultaneously. Small flowers can be done in 20 minutes and mounted using a small cover glass. Larger corollas, 2–4 cm long, may need to be soaked for hours and mounted using a larger cover glass.

These investigators used plastic ice cube trays and labels for each taxon or sample. This provides a map as to which floret and what taxon occupies each tray compartment. Solutions must be pipetted into and out of tray depressions or vials.

Entire florets, at several maturation stages, should be used in preparations: corolla, mature achenes (from which you should tease out ovules), pappus, immature ovary, style, anthers, etc. Annotate storage vials or slides with taxon and collection number.

Procedure

1. Boil florets in water. For delicate florets, place in water and make sure they sink. Florets of *Mutisieae* and *Cardueae* are larger and thicker and thus need boiling.
2. Dissect material as follows. Small florets, ca. 2 mm long, should not be dissected because anthers and styles of small florets may get lost. Dissect large florets. Rays and bilabiate (zygomorphic) florets need only be tube-sliced. Discoid (actinomorphic) florets can be longisectioned at any sinus. Slightly zygomorphic disk florets should not be longisectioned along deeper sinuses but along the more numerous, shallower sinuses. Remove style from within anther cylinder, but let styles remain attached to young ovaries. Large achenes should be split down the middle. Carefully remove ovule and also keep achene wall. Place pieces in ice cube tray compartments for subsequent steps.
3. Add diluted Clorox (The Clorox Company, Oakland, CA). Pipette dilute aqueous sodium hypochlorite to the dissections. Use 1:6 to 1:10 dilutions from the stock solution of commercial laundry bleach, which is already about a 6% solution. This will remove pigmentation and make cell patterns and venation visible. Fungus will be destroyed.

   CAUTION: Monitor progress. Too long an immersion will destroy tissues. Use very dilute solution for ca. 20 minutes for small florets. Large florets may soak for 1 or 2 hours to remove pigmentation. Put corolla and achenes in Clorox first; later add more fragile styles and anthers if they have been already separated from corollas. For small achenes, dissect out ovule after Clorox treatment.

4. Water bath to rinse out bleach, 20 seconds to overnight; time not critical.
5. Stain with safranin. Immerse corollas ca. 5 minutes.
6. Perform one to three successive water soaks, several minutes each, to rinse out excess stain.
7. Arrange stained sections or whole mounts on a slide. (a) For small florets: add ca. 2 drops of glycerin and add small cover glass. (b) For larger florets: if necessary, further dissect in drop of water on one end of slide and then add large cover glass. (c) For longi-sliced cylindrical anthers: place one half to the left, introrse side up, and one half to the right, extrorse side up. Keep filaments attached to corolla if possible.

   Carefully arrange corolla, style, achene, etc., on one end of slide with thicker parts (usually achene) to one end, not the middle, of the preparation. Cautiously, use a piece of filter paper to blot out excess water. Do not let material dry out. Immediately add 5–10 drops of glycerin.
8. Drop a correctly sized cover glass onto the thin end of the preparation, then gradually lower it onto the thicker end. Tap out air bubbles from center toward the thicker end. Seal cover glass using slow- (not rapid-) drying clear nail polish.
9. Store labeled slides horizontally.

Tissue Printing

Reid et al. (1992) published a collection on this and related topics. See especially Varner: Historical aspects, pp. 1–4; Taylor: Tissue printing demonstration, pp. 5–7; Reid, Castelloe, and Varner: Physical tissue prints, pp. 9–22.

Membranes such as nitrocellulose, Nytran (GE Healthcare, Little Chalfort, UK), GeneScreen (PerkinElmer, Inc., Boston, MA), and Immobilon (EMD Millipore, Billerica, MA) have been developed to bind proteins and amino acids, as well as other cellular structures. Researchers interested in nucleic acids and proteins can probe tissue prints using immunochemistry techniques, as they cause little or no lateral movement of substances within the membrane.

These materials can also used for making mechanical impressions of plant tissues that can be viewed directly by bright field or ultraviolet light microscopy, or by using side or dark field illumination. Depending on the hardness of the tissues, plant cell walls will cut into the membranes, leaving a permanent bas-relief impression (tissue print).

Tissue printing allows the researcher to study anatomical details that are often thought to require fixing, embedding, sectioning, and staining. A surprising amount of detail can be viewed, and the membrane preparations remain stable for long periods. There are various ways this can be accomplished, using softer membranes or harder plastic, depending on the resistance of the plant organ to be printed.

Using Membranes

1. Using a fresh double-edged razor blade, and an LDPE cutting surface, cut a tissue section of, for example, a petiole or stem, 0.2–2 mm thick.
2. Remove any excess exudates from cut surfaces by blotting with a piece of filter paper.
3. Use forceps to transfer the section, or several successive sections, to a piece of membrane.
4. Place a nonabsorbent paper over the section to prevent fingerprints, and press on the section with a finger or thumb. By trial, use just enough pressure to make an impression of the organ's cellular structure onto the membrane.
5. Carefully remove the section from the membrane and dry the printed membrane in warm air. Observe with white or ultraviolet light. Side illumination is especially informative. Varner used a dissection microscope capable of substage dark field illumination. Results are often quite striking.

Using Cellulose Acetate

Harder plant tissues (lignified, silicified, or cutinized) print well on many plastic surfaces, including the dried emulsion surfaces of photographic film, even on the cellulose acetate (reverse) side of film (the film should be unexposed and developed). However, softer tissues work less well on harder plastic membranes.

Performing Chemical Tests on Tissue Prints

If, for instance, you want to investigate a stem cross section for its pattern of starch deposition, place a drop of dilute iodine-potassium iodide solution on the membrane print. The starch pattern will become immediately evident. Its deposition pattern will be easily seen, e.g., as uniform, or only in the pith or cortex.

On tissue prints where microchemical tests are performed, the resolution of detail on membranes sometimes suffers. Higher resolution of tissue print detail has been accomplished when an agarose mixture is used. This softer printing medium is made as follows.

Procedure

1. In an Erlenmeyer flask, combine agarose 3 g; glycerol 2.5 g; sorbitol 5 g; and distilled water to a final volume of 50 ml.
2. Warm and stir until components are dissolved.
3. Heat to just below boiling to remove most air bubbles.
4. Add a few drops to microscope slide and tilt so that solution will cover ca. ⅔ of the slide.
5. Hold slide vertically to drain excess solution.
6. For best impressions, store slides for several days before use.

Solution thickness can be varied by letting the solution stand on the slide horizontally at room temperature for several minutes before draining, or by heating the slide gently to thin the solution. Ingredient concentrations can also be varied.

For Use in Microscopic Study

1. Make a razorblade section ca. 0.5 mm thick.
2. Place the section on a coated slide.
3. Place a piece of clear adhesive tape over the section and press the tape, rolling a finger side to side to assure even pressure.
4. Pull away tape with section.

The resulting impression will remain stable for at least several weeks. For best viewing use dark field illumination, or at least oblique illumination (see chapter 8).

Anatomical Fixatives

For light microscopy (LM) investigations, general anatomical fixative formulas should provide quick killing and pretty good cytological preservation. Criteria include balancing swelling against shrinkage, and hardening against softening. Alcoholic solutions penetrate tissues better than aqueous-based ones.

When making thin sections for LM, anatomists are interested in being able to detect cell and tissue types, as well as their undistorted relationship to each other. In well-fixed material, one should be able to see what types of living contents, if any, were present at the time of fixation.

NOTE: Those who wish to observe chromosome cytology or do transmission electron microscopy (TEM) investigations of cytoplasm will need more sophisticated chemistry.

General fixatives. The best general fixatives are formalin acetic acid alcohol (FAA) or formalin propionic acid alcohol (FPA). There seems to be no difference in efficacy. Proportions: formalin 5 parts; acetic, or propionic, acid 5 parts; 50% ethanol 90 parts. Ethanol at 70% strength is not necessary at this stage. After 48 hours, or weeks later, specimens should be stored in straight 70% ethanol. A little glycerin, 1%, or a squirt from a pipette (exact concentration is immaterial) will prevent specimens from drying out if the volatiles evaporate in storage.

Other published formulas as listed below will work for storage of wet-preserved plant material. They are not as good as the above for fixation; see also Bridson and Forman (1992).

Kew mixture. Industrial methylated spirits (IMS) (=denatured ethanol), formalin, glycerin, water: 10:1:1:8 (v/v).

NOTE: Stern (2004) found that although Kew mixture may be safer, it fixes poorly and tends to turn tissues to mush.

Copenhagen mixture. IMS: glycerol: water. 10:1:8 (v/v), IMS or etOH alone: 50–70% strength.

Field collections. When collecting in remote field sites, other liquids can substitute for better fixatives. For anatomical investigations, liquid preservation in a variety of alcoholic solutions usually yields better anatomical preparations than dried specimens. I've made good slides from specimens initially stored in denatured alcohol, ethanol, isopropyl alcohol, rubbing alcohol, gin, whisky, wine, etc. Note that alcohols weaker than 50% aqueous (100 proof) don't fix as well.

For field use, there are two specimen storage techniques that work better than using individual glass or plastic bottles: (a) Plastic bags called Whirl-Paks (Nasco, Fort Atkinson, WI) are liquid-proof and take up little space. Before flying back from the field, empty out the fixative, leaving a piece of wet paper towel in the bag, and re-seal; or (b) Frequently, when working from a vehicle, I have taken the plant pieces, wrapped them with a label in bandage gauze, and tied the little packages with dental floss. Such accumulated samples are then placed in a gallon-sized plastic jar, half filled with fixative. Either way, back in the laboratory, it is easy to repackage the specimens in individual bottles with fresh alcohol, usually 70% strength.

Fixative timing. Hewlett (2002) reported on fixative penetration rates, which bear on the time it takes for tissues to be considered adequately fixed. He noted Medawar's application of diffusion laws to fixation penetration. That is, depth of penetration is proportional to the square root of time. Each fixative has a coefficient of diffusibility (K), the Medawar constant. Using $d=K\sqrt{t}$, where *d* is the distance penetrated in millimeters and *t* is time in hours, *K* is the constant for fixative in question. K=3.6 to 5.5 for formaldehyde. It follows that large samples are penetrated more slowly than large ones, and that rate slows with time.

This idea may not be all that helpful. Binding may take longer following penetration since a "25-hour fixation" may actually take up to 7 days for complete irreversible cross linking. The presence of mucilage or latex will also slow penetration.

Formaldehyde is thought to fix by addition and cross linking (at least with protein) rather than by coagulation. Water and alcohol can readily reverse these reactions, perhaps as much as 90%. Also, in dilute aqueous solutions, formaldehyde readily becomes hydrated to form methylene hydrate (methylene glycol). Less than 0.1% of true formaldehyde may be present. As methylene glycol penetrates tissue, more formaldehyde is generated. While the tissue perimeter may be adequately cross linked, the remainder of the tissue is fixed by coagulant alcohol during processing. This could compromise results regardless of the size of the specimen. See Fox et al. (1985) and Kiernan (1990) for detailed discussions of the properties of formaldehyde.

Curation of Liquid Preserved Specimens

Schmid (1981) called attention to a major problem with collections of liquid preserved plants: the likely evaporation of fluid over time. Adding a milliliter of glycerin to each bottle will prevent the specimens from drying out completely. Even easier is Schmid's solution. He suggested lining lids with plastic sheeting. After experimenting with polyethylene sheeting of various weights, he concluded that lining lids with pieces of medium-weight plastic works best. Collection bag weight is too heavy. Supermarket bag pieces work if two layers are used. Personally, I have found 4-ml builder's plastic to work well. Parafilm (Pechiny Plastic Packaging Co., Chicago, IL) works, but plastic pieces are cheaper. Also, in my experience, a good plastic lid liner, not the vinyl-coated cardboard type, will cause bottles to hold their liquid indefinitely.

CHAPTER 2

Wood, Woody Stems, and Bark

In addition to the making of wood sections, this chapter also covers macerations, which frequently yield informative data on cell wall and pitting patterns.

Sectioning Wood or Woody Stems

While the sliding (sledge) microtome is the standard tool for sectioning wood samples of uniform thickness, the suggestions offered below will also provide informative sections. Sections approximating 25 μm thick are considered optimal for many species.

Trimming Wood Specimens To Be Sectioned

Traditionally, wood anatomical data are gathered from three faces or aspects: (a) Cross (transverse) sections, and longitudinal sections of two types: (b) Radial, and (c) Tangential. These three sections are usually mounted side by side beneath one cover slip and the slide labeled *x.r.t* along with the species name and collection number.

Using a small vise, or a clamp that will hold a larger piece of wood against a bench top, cut off your study sample with a hack saw, coping saw, or jeweler's saw. With larger pieces of wood, the cross section is obtained first from one end of your sawed piece. Then, radial and tangential sections are prepared by splitting with a large knife or chisel and small mallet. These splits follow the grain for better radial and tangential orientations.

For microtoming, the frequently recommended size is about 1 cm^2 for the cross section, by about 2 cm long on the longitudinal dimension. You have to size your specimen on the basis of your cutting technique as discussed below.

The end of the sample block that yields cross sections should be at right angles (orthogonal) to the vertical grain. First make your cross sections, then inspect the cut surface, if necessary, to ascertain the best trim for radial sections. Finally, make a split to obtain the tangential sections.

For small, difficult-to-handle pieces, use a support block and the superglue technique described below, under "Superglue to Support Small Plant Organs for Hand-Sectioning".

Softening

For most wood species, boil the ready-to-section sample in a beaker of water sufficient to cover the sample. Use a hot plate and heat until the sample becomes waterlogged and sinks. Alternatively, for some specimens, it is possible to get good sections by soaking the wood in a dilute detergent solution for several days. Soak out the detergent if you intend to use the superglue technique below.

Jansen et al. (1998) recommended the use of glycerin solution to soften wood. Trim the wood piece to expose cross, radial, and tangential faces. Immerse the piece in a boiling mixture of glycerin and water (1:10 v/v) until it becomes saturated and sinks. Longer boiling increases softening. The process may be accelerated by transferring the sample block to cold water for a few seconds intermittently during the boiling period. Keep water level at original mark by adding hot water. If the concentration of glycerin increases, the boiling point also increases, which could damage the sample. After this process, sectioning is often best accomplished while the wood is still hot and swollen.

Jansen notes that if the sections are to be stained, it is often preferable to bleach them first. Sections, especially thicker ones, will be rendered somewhat clearer after the stain process. Use commercial bleach (which is ca. 6% aqueous) or up to 10% sodium hypochlorite for 1–3 minutes. Rinse several times. Jansen's detailed paper should be consulted for numerous other suggestions and refinements to this process.

(Kuchachka, 1977; Carlquist, 1982). Ethylenediamine (EDA) was introduced originally by Kuchachka for softening hard woods prior to sliding microtome sectioning. As used by Carlquist, EDA can be used to soften diverse plant materials, especially those with alternating hard and soft elements. Such tissue, including wood, can be softened and then embedded in paraffin for rotary microtoming, since boiled or pickled wood is still too hard for rotary microtoming. The use of EDA can make this possible.

IN USE: Soak pickled or boiled plant material in 10% EDA, 3 days at room temperature, or 4% EDA for 1 week. Make 2 or 3 changes of water at 2-hour intervals. EDA darkens to brown in use. If this is not a drawback, solution can be saved and reused.

NOTE: The samples can be infiltrated in paraffin as usual. If wood, they should be as thin as possible for best infiltration. After embedment, trim the surface and soak the cut edge in water, and sectioning should be very satisfactory.

SAFETY: Use EDA in a hood. When unsealed, it gives off smoky fumes. It is less dangerous when diluted, but still use the hood. Wash exposed skin promptly. It can cause skin irritation.

Levings (unpublished) found that hard woods can be treated successfully as follows. Take wet wood, from a 25% ethanol soak, or previous water-soak, and trim it to the smallest piece from which you will section later. Place it in 10% aqueous EDA. Heat in oven at 26–32°C, or on a warming tray, for about 7 days. Levings found a cold soak less effective. Give two rinses in water, then boil sample until it sinks, or about 30 minutes. After cooling, cut wood sections directly or glue to a holding block, with superglue, as required. Specimens should cut as would normally dense wood.

Sectioning

Cutting Board Technique

Take a plastic disk such as a food container lid to use as an anvil or cutting board. The disk should be of low-density polyethylene (number

4 LDPE plastic), which will not damage a blade edge. Using a single-edged razor blade, trim the wood specimen by cutting downward against the plastic disc to produce wood slivers. Cut several from each x.r.t. plane and place them in Syracuse dishes, depressions in a spot plate, or watch glasses. Make wet mounts directly after staining (see chapter 6 for options). Almost always, I recover several slices that yield the necessary microscopic information.

Cutting with the Hand-held (Cylinder) Microtome

This tool is still available from biological supply companies for $40 to $300. Possible sources are euromax.eu (The Netherlands), optics-shop.com, and various internet auction sites. These sources, plus fishersci.com, also supply disposable blades and holders for disposable blades. The microtome sold by optics-shop.com, made by Perret Opticiens (Geneva, Switzerland), has a clamp that fastens the microtome to a table or lab bench.

After some practice, the cylinder microtome will yield good results when cutting many botanical materials. It stands about 8 cm tall and wide. Its top surface is a flat disk or stage, with a hole in the center fitted

FIGURE 3. Cylinder microtome. A specimen is clamped in place using the knob on the side. A blade is brought across the flat surface to make a section. After each cut, the bottom knob is turned a calibrated amount to raise the specimen for the next section. A small specimen can be supported by sandwiching it, using pieces of Styrofoam or a carrot piece bored with an appropriately sized cork borer.

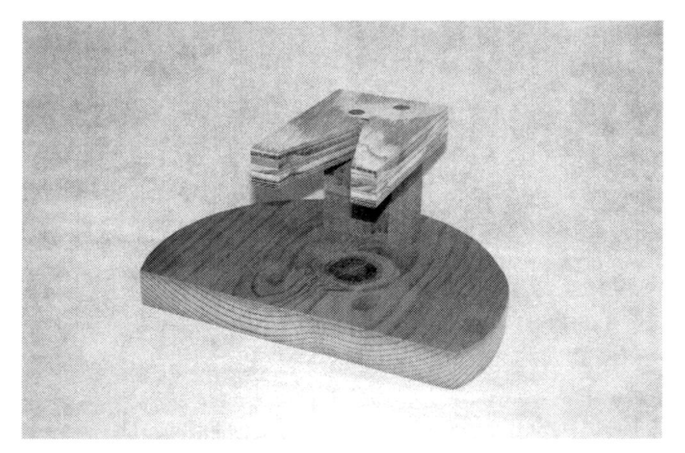

FIGURE 4. Wood frame used to hold cylinder microtome while knife is pushed across the specimen. The base of the wood frame can be clamped to the table top.

FIGURE 5. Wood frame supporting the microtome. The frame can be clamped to the table-top for support.

with a clamp. The disk's support stem is fitted with a screw micrometer on the bottom end, which allows one to elevate the clamped specimen above the disk surface a known amount between sections. Holding the microtome stem with one hand, your blade is drawn across the smooth top surface of the disk and the specimen in a slicing action. Hold the blade at about a 25-degree elevated angle with the cutting edge flush with the top of the disk. Try softer stems first, but later it will become practical to get good sections from harder wood.

Blades

Regular sharpened microtome blades, 160–250 mm long, represent the gold standard for good sectioning. The more massive the steel, the more stable the blade. If you can obtain one, it will have to be periodically sent out to a professional facility for sharpening.

(See also Netnotes column [2005], discussion by Sherman, White, and Cougar). Pieces of carrot or high-density foam can be used for support for small stem sections. A razor blade is suggested. Better, says Cougar, use a woodworker's plane iron, which can be rendered as sharp as a microtome blade. Cougar noted that a polished edge can be put on either a plane iron or wide wood chisel. Woodworker's sharpening guides can help with this for the do-it-yourselfer.

On the small microtome platform, cementing a glass slide to each side of the specimen hole can assist in smoother cutting. Razor blades or straight razors are not stiff enough for many wood samples. Use a slicing motion.

There are packages available of disposable microtome blades that are made of heavier steel than single-edged razor blades. Some brands are quite sharp and are worth trying with your material. Try Personna (Personna Blades, Verona, VA) stainless steel surgical prep blades. Thinner shaving blades are more likely to deflect during the cut, or to dull quickly. Using the cutting board technique, I have found that single-edged utility blades work adequately for all but the hardest woods.

The straight razor. The following was reported to me by the late Prof. Harry Muegel, whose mentor was the eminent Yale wood anatomist, Samuel Record. Holding a wood piece in one hand, Prof. Record drew a freshly stropped straight razor across his hand-held sample. His skill with this tool always produced sections that were quite satisfactory for diagnostic purposes.

I advise against doing it this way. If you must try this, hold the specimen while wearing a wood-carver's glove that is designed to prevent cutting yourself. These are made of steel mesh or Kevlar and are available in wood-carver's catalogs. Also available is protective finger guard tape. You're on your own!

Superglue to Support Small Plant Organs for Hand-Sectioning

(Keating and Levings, unpublished, March 2009). Occasionally structural information needs to be harvested from a small piece of wood, such as a piece of core sample, twig, bark, or other plant material. It is unsafe to attempt to hand-hold too small a sample while shaving it with a razor blade. Also, if you have use of a sliding microtome, clamping a small piece with very little clearance between the blade and steel clamp can lead to inadvertent blade-edge destruction.

The following technique, known to users of the sliding microtome, works reliably. Fashion a handle or holding block that you will use to stabilize your sample while cutting it. Make this any size you feel comfortable holding onto while sectioning, or the size that will fit the clamp of the cylinder microtome. Rather than conifer wood, which may have resin that interferes with a good glue bond, I prefer light hardwood pieces such as cottonwood, willow, or pieces of ½″ or ⅝″ hardwood dowel.

Procedure

1. As usual, for optimal cutting, wood samples are simmered in water on a hot plate until they sink. Also soak, at least briefly, the wood holding block.
2. With a paper towel or blotting paper, blot both pieces to dampness. They should never be dripping wet.
3. Using a tube of superglue (cyanoacrylate), attach the specimen to the block, oriented for optimal trimming and cutting. This type of glue operates best in the presence of moisture. Within about 5 minutes, the glue hardens. To be certain, put the assembly aside for about 20 minutes.
4. If desired, the small sample can be sliced freehand or held down for sectioning against a polyethylene (number 4 LDPE) plastic lid.

NOTES:

1. After cutting in the preferred dimension, the sample can be easily pried away from the holding block using a chisel, reoriented or trimmed, and glued down again for further sectioning.

2. When cutting by hand against a plastic cutting board, my holding block is about 10 cm long by 15 mm diameter, which I grip while hand-trimming and sectioning. The small specimens to be investigated, some as small as 5 mm on a side, are cut from twigs, or from small-diameter core samples.
3. Superglue must be stored in anhydrous condition. After first opening and use, the remaining glue may harden in the tube. Obtain a package of several small tubes and have a fresh unopened one available.

Wood Sections Cut Using a Drill Press and Forstner Bit

If your wood specimen is large enough to be clamped or held firmly on the stage of a drill press, I have found good diagnostic results with the following. Place a sharp ½″, ⅝″, or ¾″ Forstner bit in the drill press. Speed of rotation seems not to be important. Bring the spinning bit down into the specimen at a rate that creates shavings rather than dust. Retrieve and mount the shavings on a slide as usual. Samples should be boiled first, but this actually works with dry wood of some species.

Wood Samples From an Increment Corer

Wilson (1956) suggested using an increment borer to make a core sample, then make sections. Use the attachment to a handle block using cyanoacrylate glue method as suggested by Keating and Levings, above.

Maceration Methods

Although usually done with wood for measurement of xylem elements, macerations can be used on other organs as well to gather observations on cellular morphology and wall pitting.

NOTE: If the same pipettes and vials are to be used for several or many samples, they must be scrupulously cleaned between uses to avoid carryover.

Maceration Fluids

Foster (1949) recommended Jeffrey's method. Place wood slivers (often trimmed from larger samples to be sectioned) in a capped vial with 10% chromic acid, 10% nitric acid (1:1). Place in oven at 30–40°C until material becomes soft or mushy. Boiling wood slivers first before adding the acids accelerates the process.

Wash material in distilled H_2O and store in 50% ethanol for future study. Materials can be viewed unstained or stained in various lignin stains. Foster recommends staining macerated material in Safranin O, rinsing, and placing the material on a slide. Separate cells by teasing the larger pieces apart. Place on warming tray to remove excess water, and mount in glycerin jelly. Cover glasses can be sealed.

In some laboratories, the macerated cells are stained in 1% aqueous safranin, or one can use 15% ethanol as the solvent. Personally, I use 0.1%–0.5% safranin, which is thin enough to be translucent but is just as effective.

Procedure

This is a gentler method for soft tissue, such as parenchyma and leaf mesophyll.

1. Place portions of tissue in acid-alcohol: 70% ethanol and conc. HCl (3:1).
2. Remove air with a vacuum pump or aspirator.
3. Add fresh mixture to the sample and let stand 24 hours.
4. Wash thoroughly in water, and transfer to 0.5% aqueous ammonium oxalate.
5. Within a few days or sooner, soft tissues can be readily dissociated with teasing needles. This works well for isolating idioblastic ramified sclereids.

Schmid's modifications. Schmid (1982) used Jeffrey's solution in concentrations as given above. He noted that weaker solutions, such as 5% strength for each component, can be used for more delicate materials. Although suggested by some authors, it is not necessary to use freshly mixed macerating solution. Schmid used a 2-year-old mixture with no noticeable loss of action. See his paper for a careful review of all modifications of this technique and several other maceration techniques, such as Wilson and Shutts (1957) below.

Schmid noted that it is not necessary to render specimens into thin slivers as is often suggested. Use larger pieces of wood or twigs up to pencil-diameter. Scrape off bark if present. Place dry wood in macerating fluid if a means is at hand to remove air, such as a vacuum pump or vacuum oven. Otherwise, boil wood pieces in water before placing them in macerating fluid.

Allow maceration solution to work, by observation, until surfaces of pieces are soft and crushable with a teasing needle. Use of a warming tray or paraffin oven will speed up the process, but room temperature will produce results within 10–40 hours. When finished, the core of a piece will remain intact but outer parts will be macerated. The advantage of using unslivered wood is that some of the wood cells will be at the right stage of maceration within a wide range of timing.

Rinse in several changes of distilled water until solution is colorless (at least several hours). At this stage there are three choices:

1. Dehydrate material through an ethanol series, stain the material, dehydrate through xylene or limonene, and make a resin-mounted slide; *or*
2. Sonicate the samples while in the xylene solution. Use 2–10 seconds of sonication, the time determining the amount of cells released from the chunk. Schmid used an ultrasonic cleaner with an operating frequency of 40 kHz; *or*
3. In any concentration of dehydrating solution, shake the stoppered sample vial to dissociate cells or use a mechanical shaker to dissociate cells. (Glass beads can be added to the vial for better shaking, but this can break up the cells if shaken too vigorously.) Using a dropper, place some sample slurry on a slide and add a cover glass. If done carefully, the cells will spread well and make an informative preparation.

SUGGESTION: Place the partially macerated chunk on a slide and view under a dissection microscope. Using a pair of dissecting tools, scrape off some softened tissue and spread this partially separated mixture on the slide in a drop of mountant and cover. This is an especially useful approach if the juxtaposition of cell types is of interest.

Dacar and Giannoni (2001), addressing animal diets in ecological studies, used the following method to prepare seed samples for determination. Their macerating fluid is a 17.5% $NaHCO_3$ (sodium bicarbonate) aqueous solution. Seeds, seed fragments, or dried fruit parts are soaked in this fluid for 24–36 hours. This saturates the hard outer layers and allows their separation using forceps and scalpel with the aid of a binocular dissecting scope. Each seed part can be cleaned in 50% household bleach (sodium hypochlorite) for 20 minutes, washed thoroughly, and mounted in pure glycerin. A cover glass is carefully lowered in place and the mount ringed with clear nail polish.

The authors demonstrated superior clarity of cell-wall visibility when compared to preparations made with Jeffrey's solution. Layer by layer, cell outlines and starch grains are more clearly rendered. The authors note that samples may remain in the macerating fluid for months without further damage. The technique was successful with fresh, dried, fossil, and fecal samples using this safe, easy-to-obtain chemical.

NOTE: I assume that "50% household bleach" means making a 1:1 aqueous solution of the commercial stock solution, which is about 6% sodium hypochlorite.

Pulp Production Technique

Wilson and Shutts (1957) suggested an excellent method, which resembles commercial wood pulp production. Soak plant slivers in 5% sodium hypochlorite until whitened. Then boil briefly in 3% sodium sulfite. Place softened slivers in a vial and wash with tap water. Cover completely with 4% aqueous formalin, add glass beads, and shake for a uniform suspension. Smear Haupt's adhesive or serum adhesive onto a slide and add several drops of water. Pipette a suspension of the macerated sample onto a slide and let dry on a warming plate. The authors recommended iron alum hematoxylin, but any stain technique appropriate for wood can be used. Dehydrate to limonene, and add resin and a cover glass.

NOTE: As an alternative to shaking with glass beads, one can place the softened slivers onto a slide and dissect the cells apart. This would allow one to observe the juxtaposition of cell types.

Bark Maceration

Coelho et al. (2012) macerated bark cells of *Bathysa* (Rubiaceae) as follows. Specimens were placed in a mixture of 30% H_2O_2 and acetic acid (1:1) at 60°C for 24–48 hours. The resulting individual bark cells were stained in Astra blue.

CHAPTER 3

Restoration, Softening, Clearing, and Bleaching

Traditionally, taxonomists have boiled flowers or other plant parts for ease of dissection. Alternatively, it is frequently sufficient to soak plant parts in water, with or without some heat, to which a little detergent has been added. The procedures described in this chapter have been found by their authors to provide an accurate view of the position of tissues for microscopic study. Maceration procedures, treated in chapter 2, are also useful for looking at cell structure of other organs.

Dried Specimen Restoration and Softening

Aerosol OT

Ayensu (1967) recommended Aerosol OT (Fisher Scientific, Fair Lawn, NJ), which reduces surface tension and interfacial tension. Its properties include penetration, emulsification, and dispersal. It can be used for softening embedded material for microtomy. Trim the block to expose the tissue and soak a few minutes before sectioning.

Procedure

1. Dilute 10% Aerosol OT solution 1:3 or 1:4 with distilled water. Some accounts suggest 2% Aerosol OT dissolved in 10% methanol. Soak specimens for 5 hours or longer; days of soaking will not damage the material.

2. Wash briefly to remove excess aerosol. Cut specimens as required. Mount sections in ethanol:glycerol (1:1), or continue with other procedures.

Chemically neutral, Aerosol OT is sodium dioctyl sulfosuccinate. It is compatible with microchemical tests such as iodine tests for starch, phloroglucinol for lignin, iodine-sulfuric acid for cellulose, and Sudan III and IV for fats and oils. It is nontoxic in the small quantities employed in this context and is stable to 65°C. In addition to being available as a liquid in various concentrations, it is also packaged as a wax-like solid. Cut off pieces and measure by weight.

Pohl's Solution

Pohl's solution, reprinted in Woodland (2000), was described by Richard Pohl and nicknamed "Pohlstoffe" by his graduate students at Iowa State University. The mixture proportions are: Aerosol OT 1%, distilled water 74%, and methanol 25%. Pohl reported that herbarium material is soft enough to be dissected within minutes. The higher concentration of alcohol probably accelerates penetration of the mixture into resistant plant tissue.

Contrad 70

(Schmid & Turner, 1977). Contrad 70 is sold as a biodegradable surfactant and radioactive decontaminant. Unlike the chemically neutral wetting agent Aerosol OT, Contrad 70 is alkaline. Its pH is 11–12 for a 5% solution, as compared to 14 for NaOH. The proprietary formula is said to include less than 3% KOH. The authors found this solution to be superior to other tested formulas for restoration of dried plant materials. It can be used to prepare materials for general taxonomic dissection as well as anatomical study.

Contrad 70 is used as a 2–5% aqueous solution in which dried materials are immersed for periods varying from one hour to several days. Timing is not critical. Ambient temperature is usual, although heat can be applied (in stoppered vials), up to the 60°C paraffin oven temperature. Higher solution concentrations or additional heat should only be applied in the case of restoration-resistant materials. After required softening action, Contrad 70 should be thoroughly washed from the tissue.

The authors found that Contrad 70 accomplished rapid and effective expansion of tissue and better appearance of resulting sections because of some removal of obfuscating tannins. Microtoming was easier because of better softening. It has no deleterious effect on cutin, starch, crystals, tannins, and secretory compounds.

Ammonium Hydroxide

Tillson and Bamford (1938) used this formula, which was also summarized by Venning (1954).

Procedure

1. Soak pieces to be sectioned in tap water, with a little wetting agent or detergent added, at 60°C for 8–10 hours. Use a corked shell vial or small capped bottle in a paraffin oven.
2. Change to dilute ammonia, NH_4OH 1:19 (=5%) at 60°C and leave overnight. Specimen should be well rounded or restored to natural shape. Old or more woody material may require longer immersion in both solutions.
3. Wash in running water 4–6 hours or until free of ammonia.
4. Fix as usual for fresh material. Wash in water 6–8 hours and proceed as with fresh material.

Procedure as Summarized by Garay (1979) and Toscano de Brito (1996)

1. Dried orchid pollinaria are placed in concentrated NH_4OH in a fume hood for 30 minutes to 12 hours, by inspection.
2. When reconstitution appears complete, wash specimen three times in distilled water to remove ammonia.
3. Place specimens in Copenhagen mixture, or FAA for storage.
4. Observe specimen under binocular microscope.

NOTE: It may not be possible to remove pollinaria intact from dried pressed flowers; when necessary, soak whole flowers in ammonia. Copenhagen mixture is a stable storage medium for wet specimens. For further treatment using a scanning electron microscope (SEM), an alcohol dehydration series may be used (see original reference).

Photo-Flo

Valdés-Reyna and Hatch (1995) developed this mixture and technique for studying grass leaf anatomy:

1. Leaf blades from dried specimens are placed in an aqueous solution of Kodak Photo-Flo 200 (Eastman Kodak Company, Rochester, NY) and water, 1:3. After specimens are thoroughly imbibed, transfer the leaves to 70% ethanol.
2. From here, leaf blades can be prepared for paraffin embedment or sectioned by hand.

Procedure for Study of Adaxial and Abaxial Epidermis Layers

1. Leaves are removed from the Photo-Flo solution and placed in Clorox for a few minutes to bleach the chlorophyll.
2. Place leaf piece flat on a microscope slide and scrape with a razor blade until uppermost epidermis, mesophyll, and vascular bundles are removed.
3. The remaining epidermis is stained with azo black, washed with 90% ethanol, and mounted in Euparal (ANSCO Laboratories, Manchester, England) and ethanol.

NOTE: Kodak Photo-Flo 200 is a powerful aqueous wetting agent, resembling Triton X-100 (Dow Chemical Co., Midland, MI), and is said to contain propylene glycol and p-tert-octylphenoxy polyethoxyethanol. Photo-Flo is effective as a surfactant at very small dilutions. One or two drops per pint of water is adequate for photographic use. A greater concentration than 1–2% aqueous solution, in my experience, is overkill for restoring most plant materials of other families.

Trisodium Phosphate (TSP)

Benninghoff (1947), summarized by Venning (1954), recommended trisodium phosphate (TSP) for restoring herbarium material as well as fossil plant specimens, including pollen, in peat.

Soak material in the solution at 60°C for 2 hours. Benninghoff recommended adding an additional wetting agent to the material and avoiding prolonged treatment in the solution. Restoration solution: TSP: variously 0.25% to 0.5% aqueous solution.

Van Cleave and Ross (1947) noted that TSP was used originally for reclaiming dried zoological specimens such as worms, but it bears experimenting with botanical materials. It seems to enhance permeability of tissues to histological reagents. The effect goes beyond reducing surface tension.

There is no apparent damage, as occurs when specimens soaked in sodium hydroxide or potassium hydroxide run toward maceration. It seems superior to some other detergents. With pickled or dried worms, soaking in TSP also provides subsequently more brilliant staining than with untreated specimens.

IN USE: Pickled specimens are passed down to distilled water, then to 0.25% aqueous TSP in distilled water. Only a few minutes may be necessary for fresh or pickled specimens to become more pliable and translucent. Hard or brittle specimens may require hours or days in a warming oven before becoming pliable. Liquid preserved zoological materials soften over time in TSP; if this is also true of botanical materials, TSP could be useful prior to making hand-sections of hard organs. Change to water when the degree of restoration or softening is reached.

Rao's Solution

Rao (1977) recommended a solution that is said not to deteriorate with time or repeated use: 20 ml glycerol; 10 ml acetic acid; 10 ml EDTA (0.292% aqueous); 10 ml sodium lauryl sulfate (5% aqueous); and 50 ml distilled water.

Soak material for a period depending on its hardness. No heating is required. Flowers can be dissected and retained in same solution with no drying or rotting, or they may be washed and dried and replaced in herbarium. To prepare for free-hand sectioning, first rinse specimens thoroughly with distilled water.

Ethylenedinitrilo Tetraacetic acid disodium salt (EDTA) is said to chelate divalent metal ions from the middle lamella and cell walls, thereby softening the material. Acetic acid preserves without hardening; sodium lauryl sulfate is the wetting agent. Glycerol softens.

Hydrofluoric Acid

Vega et al. (2008) used hydrofluoric acid to desilicify grass leaves before embedding for paraffin sectioning. It may also provide better hand-sections of tough grass tissues. Using liquid preserved material, or dried material softened in 5% Contrad 70, material was desilicified in 5% HF for 24 hours, then washed in distilled water.

NOTE: HF is very corrosive, must be kept off skin, and cannot be stored in glass containers. It must be stored and used in a hood.

Clearing

Sodium Hydroxide

Foster (1949) recommended the following for viewing venation or sclereids. Clear small flowers, whole leaves, or leaf pieces using 5–10% NaOH. If dry, soak leaf first in hot water until it sinks. If fresh, extract chlorophyll with hot alcohol. Thin, lightly colored material can be treated at room temperature. More resistant or coriaceous leaves should be placed in an oven. Change reagent if it becomes dark colored. When translucent, the leaf can be studied at low power. It can be stored in 50% ethanol for future observation.

If still opaque, dehydration in an ethanol series and resin-mounting will often render venation clearly seen. Thick mounts will take up to a week on an electric slide warmer to be handled safely, since enough solvent needs to be driven off before resin will thicken sufficiently. Stains such as safranin or Delafield hematoxylin may be used. For small flowers, using depression slides, or Raj slides, will increase the space beneath the cover glass for a thick mount. See description of Herr's slides at the end of this chapter.

Morley (1949) noted that plant materials cleared on NaOH frequently have less affinity for dyes. Sometimes tissues can be observed microscopically without staining, but frequently, cleared, aqueous, unstained specimens have a milky appearance. A dehydration sequence ending in xylene, toluene, or limonene, plus resin-mounting, renders the specimens transparent. Because the specimen's RI becomes similar to that of glass, the desired structures may become nearly invisible.

Except for some techniques summarized elsewhere in this manual, Morley noted difficulties with many popular dyes when used with cleared material. These include the tendency of dyes to leach into the mountant, as well as lack of sharp differentiation. In addition, different genera from successful original tests may not stain well at all.

Morley screened more than 20 dyes and ranked them in terms of their ability to differentiate features of epidermis, mesophyll, sclerenchyma, and tracheary elements. His most successful results were accomplished as follows: (a) For epidermis, $ZnCl_2$-tannic acid-iron alum. Use 1% tannic acid (Johansen, 1940, p. 85); (b) For mesophyll, Foster's tannic acid-iron chloride (Johansen, 1940, p. 91); (c) For sclerenchyma and tracheary elements, Safranin O at about 0.5% concentration in xylene and 100% ethanol (1:1 v/v).

As a rule, overstaining can often be solved using acidified alcohol rinses. Conversely, if too much destaining has occurred, replace the specimen in the dye. Experiment with many dyes and consult Morley's paper for detailed discussion.

Sodium Hydroxide and Chloral Hydrate

Arnott (1959) suggested the following technique, which has become widely used.

Procedure

1. Place leaves in 5% NaOH at 37°C, one to many days, until almost all pigment is removed from the leaf and brownish tinge remains.
2. Water rinse.

3. Place in chloral hydrate (50 g to 30 ml water) one to many days.
4. Water rinse overnight.
5. Add 70% ethanol. Stain in 1% Safranin O in 50% ethanol one minute, then the following changes: 70% ethanol one minute, 95% ethanol one to several minutes, abs alc/xylene one to several minutes, xylene at least 10 minutes, then mount in resin.

NOTES:

1. Change solutions with a syringe as leaves become very soft during the clearing process. Leaves become very transparent in chloral hydrate and may become cloudy during water rinses. Clarity is restored during dehydration.
2. My results produced poor differentiation. Veins are fairly clear and red, but against a pink background. Chloral hydrate has a high RI that yields a glass-clear background with some materials. Its problem is that it tends to destroy stain differentiation.

Stockwell's Bleach

(Johansen, 1940, p. 85; Schmid, 1977). Plant organs fixed in standard fixatives, and even dried material, often develop dark coloration that interferes with either staining, or viewing the cellular structure of unstained sections. Especially common in certain families, the coloration is usually due to phenolic compounds, tannins, tannin precursors (phlobaphenes), or other oxidative products. Stockwell's solution is especially useful when usual clearing techniques work too slowly or fail. Mix 1g chromic acid (chromium trioxide), 10 ml glacial acetic acid, 1g potassium dichromate, and 90 ml water.

Transfer fixed materials or restored materials to water, then transfer to Stockwell's solution. Leave 20–30 hours; actual time by observation. When left too long, some tissues become fragile and hard to handle. Transfer to water. Dehydrate or process the usual way.

For embryos, dehydrate to 100% ethanol, and transfer to 4½ or BB-4½ clearing fluids (Herr, 1982). This method is described in chapter 5.

Stockwell's Bleach plus Lactic Acid/Chloral Hydrate

Morawetz (2013) developed a clearing method for Orobanchaceae haustoria that provided the best observation of the organ's vascular tissue. The selected haustoria had especially darkly stained tissues, and this technique greatly improved their translucency.

Procedure

1. Specimens were taken from 70% ethanolic solution and bleached in Stockwell's for 3–10 days until tissues were opaque-white. Bleach was changed daily or when darkened.
2. Tissues were rinsed in water 3 times, then dehydrated in a graded series of ethanol up to 100%.
3. Tissues were placed in lactic acid/chloral hydrate solution at 42°C until cleared, usually about 3 days.
4. Tissues were washed in 100% ethanol three times, rehydrated, then stored in 70% ethanol.
5. Specimens were photographed using transmitted light microscopy.

Clearing and Staining: Leaves and Other Organs

Lersten (1967) collected useful references from the scattered literature on this topic. Also, see collections by Bersier and Bocquet (1960) and Gardner (1975). Many workers have found one or more of these techniques to be quite useful for observing features such as tissue patterns, vasculature, crystals, sclereids, and other gross structures.

The tissues are rendered sufficiently transparent or translucent with reagents that remove coloration. In the process, cytological information is largely destroyed. O'Brien and McCully (1981) explain in detail what happens during clearing.

NOTE: In many of the following procedures, specimens become quite soft at times, and one should avoid disturbing them while they are being processed. Leave the specimen in its vessel and change solutions with a pipette, or gentle decanting if possible.

Clearing with Papain

Rodin and Davis (1967) noted that some liquid-preserved specimens do not clear well in NaOH or other clearing agents. After washing FAA-fixed specimens in running water for 3 days, the specimens were incubated at 40°C for 5–7 days in the following mixture: 2% papain buffered to pH 7.2 and activated with 15 ml of 0.02M Na_2S. The partially digested cell contents were completely removed when clearing was resumed using NaOH followed by chloral hydrate. Staining was successful using Safranin O, 1% in 50% ethanol for 12 hours. This was followed by dehydration, clearing in xylene and resin-mounting.

Simultaneous Clearing and Staining

Fuchs (1963) added basic fuchsin dye directly to clearing solutions: Only phenol-rich sites remain to be stained (see Schiff's reagent: Jensen, 1962). Fuchs found vascular elements to be differentially stained and made more visible than with other methods tried. He recommended modifications of the procedure for softer and harder plant materials.

Method 1 (for soft materials)

1. First, prepare the dye/NaOH mixture: 1 g basic fuchsin dissolved in 100 ml boiling water. Do not filter. When cool, add 10 g solid NaOH.
2. Place samples in 80% ethanol several days.
3. Rehydrate slowly.
4. Place samples in dye/NaOH mixture, 10–14 hours at 60°C.
5. Wash in water 12 hours with frequent changes. The sample should be a grayish color with red lignified elements.
6. Dehydrate for ca. 12 hours in an alcohol series (50%, 70%, 95% ethanol). Replace each solution with next higher concentration of ethanol when no longer colored. Transfer to absolute ethanol 1 or 2 hours.
7. Place in mixture of absolute ethanol: concentrated HCl (3:1). The red lignified tissues become dark green. Leave in mixture until other tissues become bleached, 1–15 minutes.
8. Wash in absolute ethanol several times over 24 hours.

9. Pass through 2 xylene baths. When cleared, mount in resin.

Method 2 (for harder materials, e.g., *Rhododendron* leaves or *Pinus* needles)

1. Place samples in 95% ethanol several days.
2. Stain for 24 hours in 1% basic fuchsin in 95% ethanol.
3. Wash in water until completely rehydrated.
4. Place sample in 15% NaOH at 60°C until completely cleared (pale yellow). As needed, renew NaOH every 24 hours. Clearing may take several days.
5. Proceed from step 5 as in method 1.

Method 3 (alternative method for harder materials)

1. Place specimens in 95% ethanol for several days.
2. Rehydrate progressively.
3. Place in 15% aqueous NaOH solution, at 60°C until cleared completely.
4. Stain in dye/NaOH mixture as in method 1, but with 15 g of NaOH.
5. Wash and complete procedure as in method 1, steps 5–9.

NOTE: During processing, it is preferable to leave the specimen in one dish and change the solutions by pipette. Methods 2 and 3 can also be used with softer materials but using 10% NaOH instead of 15%. If the final clearing seems inadequate, use water rinses acidified with 5% HCl, which will intensify the red of the lignified elements.

(Kurth, Keating, unpublished). The following technique was developed using *Scolopia* leaves.

Procedure

1. Place specimen in 2% Aerosol OT until uniformly wetted. Rinse.
2. To begin clearing, add 5–10% NaOH and place on a slide warming tray 24 hours. This is too short a time for clearing and the leaf will still be mostly too dark.

3. Add safranin stain for several minutes, which further darkens the leaf. We used Johansen's safranin, but other formulas will work. Rinse.
4. Add fresh NaOH until cleared to a milky off-white.
5. Dehydrate through a t-butanol series to butyl IV, abs alc, xyl/alc, xylene, then mount in resin. Results are better than most stained, cleared sections. Safranin remained in the veins but not the ground tissue.

NOTE: Not yet tried: mount in calcium chloride solution following step 4.

Clearing with Heat

Dry Oven

(Keating). It is better to use some heat since room temperature treatments will take longer and increase the risk of specimen maceration. Without an oven of any type, it is possible to use an incandescent light bulb within a fiberboard or wooden "heat box." In my test, a 100-watt tungsten bulb, in a 1-ft^3 plywood box, raised the temperature to 46°C, which is enough to hasten the clearing process. I mounted a porcelain socket on a wood block and arranged to keep the bulb from touching any surface. To avoid possible fire, be scrupulous about bulb clearances.

Procedure

1. Pretreat leaf with (a) Photo-Flo:water (1:3); or (b) Parson's Ammonia (Church & Dwight Co., Inc., Princeton, NJ) or other dilute commercial ammonia cleaner. Leave specimen in heat box for 2 to several hours. Skipping this step could result in uneven or blotchy clearing.
2. Change to 5% NaOH or KOH and leave in heat box for 12–14 hours. Specimens should become light tan.
3. Rinse specimens in about two changes of water.
4. Add Clorox; at room temperature, leave specimen for about 2 hours or until leaves are uniformly white or off-white.

5. Rinse in several changes of water by pipetting the solutions from the now soft specimens. In one change. Let specimen soak for an hour.
6. Add I_2KI solution and let soak for about 4 hours. It can be prepared in one of two ways, the results being identical: (a) I_2KI is mixed 2% strength in 15% ethanol; or (b) I_2KI is added directly to 20% $CaCl_2$ (v/v). Specimens will appear dark amber.
7. Without rinsing, mount specimen in 20–30% $CaCl_2$. The specimen doesn't look very transparent at first. After 12 hours, the results are outstanding. Even fine veinlets are bright and contrasted against a glass-clear or slightly amber background.

NOTE: Species tested in this protocol varied somewhat in the times necessary for each step.

Autoclave or Pressure Cooker

(O'Brien & Teichman, 1974; see also O'Brien & McCully, 1981). The authors suggested using an autoclave or pressure cooker. Leaves are placed in 70–80% lactic acid, in an autoclave-resistant jar, and autoclaved for 15 minutes at 15 psi.

Or, one may autoclave briefly in ethanol to remove chlorophyll, then autoclave again in 1–5% NaOH, depending on the resistance of the leaf. After autoclaving, the specimens may be treated with bleach, which clears uniformly, and not just from the edges. They note that autoclaving the specimen in bleach results in tissue maceration.

Microwave Oven

(See also Schnichnes et al., 1999; Hayat, 2000; Kappe, 2001). The magnetrons of microwave ovens produce non-ionizing microwaves, usually of 2.45 GHz frequency. This energy-efficient means of heating does not rely on conduction or convection. Energy passes through the vessel and heats the reactants and solvent.

Materials dissipate microwave energy by either dipole rotation or ionic conduction. When molecules with a permanent dipole are submitted to an electric field, they become aligned. As the field oscillates, the orientation changes continuously, resulting in intense internal heating. During ionic conduction, the dissolved charged particles in the solvent oscillate back and forth in the energized field. They collide with neighboring molecules, creating heat. Such molecular agitation hastens penetration of solvents, fixatives, and clearing agents. Therefore, protocols that traditionally call for many hours or days of soaking a specimen, with or without heat, can be accomplished often in a matter of minutes within the microwave oven.

Inexpensive domestic ovens can be used for fixing, infiltration, or clearing with the understanding that there are deficiencies in temperature and pressure control, and uneven electromagnetic field distribution.

CAUTIONS:

1. Loosen the caps on specimen bottles or they might explode. (The use of pressure vessels and specialized laboratory microwave ovens is beyond the scope of this discussion.)
2. Use glass or plastic; no metal can be placed within the oven. Pencil- or ink-written labels attached to the specimen bottles cause no problems.
3. It is important not to let the specimens boil vigorously, unless maceration is the intent.

TO PREVENT BOILING:

1. When using the defrost cycle (pulsed magnetron radiation) the magnetron cycles on and off, 1–3 seconds per state. Because magnetrons are either fully on or off, the ovens can run under reduced power only by using an on/off cycle. This lengthens the time until boiling would occur; *or*
2. Better yet, when placing specimen bottles in the oven, add a vessel of water, e.g., 400 ml in a beaker, or a Pyrex glass baking dish (World Kitchen, LLC, Rosemont, IL). This constitutes a *static water load*. Microwave energy is distributed among all of the liquid in the oven cavity, and the temperature is maintained below boiling for a period based on the total volume of liquid within.

The timing required for specimen clearing, for example, depends on the magnetron wattage as well as the total water volume. Test as follows: With just water in the vials and static water load present, operate the oven on full power for test periods of several minutes. Open the oven and record the temperature that develops in the liquid for those periods and keep a chart. The time required to raise the water temperature to 70–80°C will be effective. Determine this before microwaving valuable specimens.

One-Hour Leaf Clearing

(Keating). Frequently, it is not necessary to clear an entire leaf. For whole leaves, specimen-handling and slide-mounting are more complex and require ample laboratory space for processing and storage. Usually the gross architectural features, including venation, can be ascertained with intact leaves, fresh, pickled or dried. Use a dissection microscope, with or without strong substage illumination as required. (See Ellis et al., 2009, for terminology.)

The clearing of leaf pieces will provide adequate data on epidermal cell shape and orientation; stomatal types, counts, and stomatal index; the nature of vascularization of areoles; the presence and orientation of crystals; the nature of tracheoidal veinlet endings; and the type and relationship of marginal tooth venation to general leaf architecture.

Procedure

1. Select a healthy leaf and cut out a 1–2 cm^2 piece to be cleared from the center or margin. Dried leaves are most resistant to disintegration, with liquid-preserved leaves being second best. Fresh leaves should probably be first dried or hardened in alcohol. Note that leaves from herbarium specimens that were microwaved during drying, or later for pest control, may be more brittle than desirable.
2. Place leaf pieces in a screw cap vial or small bottle and add Aerosol OT, 2% strength in 10% methanol.
3. Wait a half-hour, rinse, and add a solution of 5% KOH or NaOH. (I have found no difference in efficacy between them.) Place the vial,

or batch of vials, *with lids loosened*, in a flat-bottomed Pyrex baking dish. Surround the base of the vials with a static load of 250–300 ml water. Place the baking dish with specimen in a microwave oven. Set the dial for about 6 minutes.

NOTE: Microwave ovens vary in power and in the presence of hot or dead spots. (My oven has a 950-watt magnetron.) As noted above, experiment with a static load minus specimen. Using your test results, the microwave time should be enough to raise water temperature to 70–75°C. Do not boil the specimens. Upon removal from the oven, the hydroxide solution will be colored with leaf pigment contents and the specimens will be lighter colored and translucent.

4. Using a pipette, rinse the specimen twice *with water from the static load*. This water is the same temperature as the specimen. Drain vial, add 6% sodium hypochlorite (stock solution of commercial laundry bleach). Watch the specimens over the next 10 minutes. When specimens are light cream colored or white, draw off the bleach and rinse with two changes of water.
5. On a labeled slide, place a drop of 20–40% calcium chloride. Using a narrow spatula, lift the specimen from the rinse water, and blot the edge against a piece of paper towel. If necessary, use a magnifier to ascertain the abaxial/adaxial orientation. Place the specimen on the slide onto the drop of calcium chloride, and add another drop on top of the specimen. Place cover glass. Within 15–30 minutes the specimen will be ready for viewing.

NOTES:

1. The advantage of this shorter schedule is that the specimen remains firm and easily handled after processing. If a multi-hour soak is performed with either hydroxide or bleach, many leaves will become too soft or begin to macerate. This makes it nearly impossible to mount on a slide with intact structure.
2. Some tropical leaves, such as those of some species of mangrove species and those of *Calophyllum* or *Duabanga*, are extremely coriaceous. They tend to resist all known clearing methods and will go straight to maceration with concentrated reagents, but see Rao (1957) below.

TROUBLESHOOTING: Occasionally, during microwaving, leaf tissue (e.g., *Eugenia uniflora*) will separate within the mesophyll and form a pillow, which destroys the final photogenicity of the clearing. This is probably due to internal structural weakness coupled with air trapped within the mesophyll and incipient boiling. To counter this, try one of three things: (a) Use a larger static load quantity to reduce the temperature; (b) Reduce the time in the microwave oven; or (c) Before microwaving, place the leaf in a portable vacuum chamber while it soaks in hydroxide solution at ambient temperature. Subject it to about 25 inches of Hg vacuum for an hour, after which microwaving works flawlessly.

NOTE: Plastic bell jars that use a tubing connection to a plastic manual vacuum pump can be obtained inexpensively.

Hydroxide, Bleach, and Iodine

Keating (2000) devised this technique before using the static water load as described above. Leaves are cleared as is frequently practiced, using hydroxide and bleach. Iodine-potassium iodide is used just prior to mounting. Magenta venation is clearly visible against a clear background. The technique is problematic only if the leaf is laden with starch.

Whole leaves are placed in a Petri dish, or larger leaves are placed in appropriate containers such as Pyrex baking dishes.

Procedure

1. Leaves, pickled or dry, are placed in 5% NaOH at room temperature and then pulse-microwaved using the defrost cycle: 2 seconds on/off for three cycles of 20 seconds each. Do not let the solution boil.
2. The NaOH-covered specimens are then placed in a 45°C oven for 48 hours, by which time the specimen is cleared to a milky off-white. Give three tap-water rinses, gently, using a pipette.
3. Immerse specimens in 6% sodium hypochlorite for 30 minutes or until the specimen is white. Rinse gently in 3 changes of water. Specimen will be very soft: draw off liquid with a pipette.
4. Add I_2KI solution (iodine 0.2%, potassium iodide 2.0%, in 15% ethanol) for 2 hours or until the leaf is gray or brown-opaque.

5. Without further rinsing, transfer the specimen to a 20–30% $CaCl_2$ solution. Within minutes, dark magenta veins will emerge in contrast to an otherwise transparent background in the cleared specimen. By the next day, the veins and other lignified material will appear a brilliant magenta color. The $CaCl_2$ will not evaporate but more may be added to replace water at the cover-glass edges for a day or two. Store specimens flat, indefinitely.

NOTES:

1. Microwaved leaves tend to clear uniformly instead of irregularly blotchy. Smaller leaves may be placed in a covered Petri dish over which a plastic bag is zipped, then placed in the microwave oven. If the bag begins to puff up, the heating should be stopped immediately to avoid boiling. Reduce power as required to keep boiling from taking place.
2. I_2KI can be added to calcium chloride to eliminate a step (to approximately 5%). These slides can be ringed using nail polish or paraffin; but, because of their thickness, they require care in handling. They must be stored flat. The stained clearing itself is archival. Leaves that are stained using Safranin O, then dehydrated and mounted in resin, can certainly be handled more easily.

Hydroxide, Bleach, and Toluidine Blue O

(Keating). Although iodine-potassium iodide staining, used after bleach, is a quick and dramatic way to demonstrate leaf venation, occasionally a leaf has enough mesophyll starch that the blue-black starch reaction darkens the whole leaf and nearly obscures the venation.

There is still hope. After bleach and a couple of water rinses, instead of IKI, add Toluidine blue O (0.1–0.5% solution in 15% ethanol) and let the specimen sit a couple of hours until leaf is uniformly blue-violet. Give 2 or 3 water rinses over a 2-hour period to leach out excess dye. Transfer leaf to 20–40% calcium chloride. Only the veins will turn blue. Over a period of weeks, the blue color will fade.

Shobe and Lersten (1967) developed a technique for clearing and staining gymnosperm leaves. Their preparations produced very clear microscopic images. The following is a summary of their preferred methods, but see my notes below.

Procedure

1. For fresh leaves, remove the chlorophyll by placing leaves in 70–95% ethanol.
2. For dried or liquid-preserved leaves, omit step 1 and proceed as follows. Place specimens in 5–10% NaOH until the leaves are mostly clear. If necessary, change NaOH as it becomes discolored. This could take several days, or hasten the process in a 16°C oven. If dark areas still remain, change to full-strength chlorine bleach for 2–5 minutes. Too long in bleach can disintegrate tissues, so watch the process.
3. Change water rinses three times.
4. Change to 250% chloral hydrate for several hours. Tissues will be rendered transparent in several hours, and specimens can be stored in this solution.
5. Rinse in three changes of water. With fragile specimens, make changes through reduced concentrations of chloral hydrate to avoid specimen damage through solution currents.
6. If specimens remain very soft, dehydrate to 95% ethanol, through a graded series.
7. Stain in the Safranin O, fast green FCF sequence. It can be run with either dye first, as in A or B below.
   A. Place specimens in fast green (in 95% ethanol, dye concentration unimportant) for a few seconds; rinse in 95% ethanol; two changes of absolute ethanol; 1–5 minutes in safranin (1% in mixture of xylene/absolute ethanol [1:1], filtered before using); destain in xylene/ethanol to required intensity; transfer to pure xylene to stop destaining; mount in resin or store in vial of xylene.
   B. Place specimens in safranin (1% on 95% ethanol) for 1 minute; rinse in 95% ethanol; stain in fast green (concentration unimportant) a few seconds; dehydrate in few changes of absolute ethanol; complete required destaining in xylene/ethanol; change to xylene; mount in resin or store in vial of xylene.

NOTES:

1. As noted elsewhere, limonene, e.g., Histo-Clear (National Diagnostics, Atlanta, GA), is a much safer alternative to xylene.
2. Chloral hydrate is a regulated, hard-to-obtain drug. Furthermore, in my opinion, any tissue stains better if chloral hydrate hasn't been used. Although cleared specimens look milky after NaOH or bleach, they become transparent when later dehydrated or mounted in calcium chloride.

(Rao, 1957). To hasten clearing in *Memecylon* leaves, the author lightly rubbed the leaves with sandpaper (no grade given) before placing in 2–5% aqueous NaOH. Rinse thoroughly with water and transfer to chloral hydrate. Some specimens were stained with safranin, dehydrated, and resin-mounted. Others were teased so as to mount free sclereids, and stained with dilute cotton blue (a.k.a. aniline blue) in lactophenol.

Visikol

Villani et al. (2013) have developed a product called Visikol (Phytosys, LLC, New Brunswick, NJ) designed to replace the difficult-to-obtain chloral hydrate. This proprietary compound is a polychlorinated alcohol mixture that also contains some glycerin. The authors' tests showed an RI slightly higher than chloral hydrate. In comparisons, the new product clears and makes slide mounts as well or better than chloral hydrate. As there are no drug regulation compliance issues, the product is efficient to obtain and use. It should be especially useful in leaf clearing, although other organs should be tried as well.

Lactic Acid

(Simpson, 1929). This technique was originally devised for looking at flowers, young fruits, and succulent stems. Materials sized to fit beneath a cover glass, thick sliced or whole, are placed in capped vials or jars in 75% lactic acid. Leave in a constant temperature oven several hours to several days until clear. Use gentle heat (up to 60°C) as the lactic acid may be thickened or hardened if water is allowed to evaporate. Specimens can then be mounted, covered, and sealed to make permanent mounts. If left unsealed, lactic acid, being hygroscopic, becomes sticky.

O'Brien and McCully (1981) noted that lactic acid has an RI of 1.4. The contrast for photography is often favorable for unstained vascular elements. The technique is gentle enough to recommend it for fragile materials such as petals or very thin leaves.

Ammonium Hydroxide and Hydrogen Peroxide

(Stebbins, 1938). Originally used for flowers, this technique works well with other organs. Cell contents are removed and vasculature stands out clearly.

Procedure

1. Boil object in water for 2–3 minutes.
2. Bleach in a mixture of concentrated ammonium hydroxide and hydrogen peroxide (1:1 to 1:2) for 1–3 days. The higher the degree of oxidation needed, the greater the peroxide.
3. Transfer to 95% ethanol for 1–12 hours for hardening. From here the specimens can be further dehydrated through a n-butanol series into a resin solvent and resin-mounted.

NOTE: Wet mounts using calcium chloride solution could also be tried.

Sodium Hydroxide and Acid Fuchsin

Ellis et al. (2009) recommended this clearing technique.

Procedure

1. The leaf is placed in a glass container, such as a Petri dish, and covered with a piece of fiberglass mesh, which facilitates changing solutions.
2. Immerse leaf in 1–5% NaOH for 2–10 days, the strength dictated by the difficulty of the specimen. Change hydroxide as it darkens due to leached pigments.

3. Change to commercial bleach (sodium hypochlorite) for 5–30 seconds, then rinse in water.
4. Wash leaves in 50% ethanol, stain in 1% acid fuchsin for 3–8 minutes. Dehydrate leaves through an ethanol series of 50%, 95%, 100% solutions. Excess dye leaches into the first two solutions while the last one stops the process.
5. Rinse in clove oil (optional), xylene, then a 1:1 concentration of xylene:Hemo-De (Meridian Bioscience, Inc., Cincinnati, OH).
6. Float leaf in dish over a light source for photography (see chapter 8). Permanent resin-mounted slides can be made at this point.

NOTE: Although Safranin O was used originally for most of the reference slides used to compile Hickey's studies in leaf architecture (Hickey, 1973, 1979), Ellis and the team working on the current edition of the manual recommend acid fuchsin as the best dye.

Hoyer's Medium

Anderson (1954) recommended this mixture especially for bryophyte tissues. Chloral hydrate is the active clearing ingredient, and it has same clearing effect on vascular plant tissues if they are lightly pigmented and not tanniniferous. (This mountant is cross referenced in chapter 7.)

Mix the following ingredients, in order, at room temperature: 50 cc distilled water; 30g gum arabic (USP flake); 200 g choral hydrate; and 20 cc glycerin.

Gum arabic goes into solution slowly. Flakes are best, but crystals will work. Avoid powder as it is difficult to work with. Use any available mechanical means of mixing. The mixture may seem to need filtering, but this is not necessary. After a few hours bubbles will disappear and the solution will clear.

Procedure

1. Well-hydrated specimens of bryophytes or other small, thin plant materials should be transferred directly to drops of Hoyer's on a

slide. Lower a cover glass carefully from one side to avoid trapping bubbles. Place flat on tray or work area.

2. Check specimen over a couple of days. At the cover-glass edge, a slight excess of mountant will evaporate, slowly forming a hardened margin. Away from the edges, the mountant remains viscous, so preparations should be stored flat.
3. Poorer grades of gum arabic may make a darker solution and perhaps leave sediment. Use the purest grade possible.

NOTE: Before Anderson introduced this formula to bryologists, Hoyer's medium was considered useful for mounting mites, small insects, and fungi. The finished medium is very light yellow and has an indefinite shelf life in well-capped containers. It clears small plant parts and serves as a good permanent mountant for small unstained whole mounts. Chloral hydrate destroys general stainability, a moot issue if you can discern your target cellular outlines.

Herr's 4½ Fluid

(Herr, 1971, 1972a, 1972b). Pistils are dissected from flower buds just prior to anthesis, fixed in FPA_{50}, and stored in 70% ethanol. Herr's 4½ fluid is made as follows: Mix by weight: lactic acid, chloral hydrate, phenol, clove oil, and xylene 2:2:2:2:1.

Procedure

1. Ovules are dissected from pistils and transferred to Herr's 4½ fluid. In 24 hours, at room temperature, the ovules appear transparent.
2. With a Pasteur pipette, the ovules plus some of the clearing fluid are transferred to culture (well) slides, which have sufficient clearance to prevent the cover glasses from bearing down on the ovules. A chamber slide can be made (see below) that meets this requirement.
3. Herr used phase contrast optics to examine the ovules. Focusing through the ovules reveals the details of megasporogenesis.

An improvement on this method, given later in the same reference, greatly enhances the phase contrast image of the megagametophyte.

Procedure

1. Treat dissected ovules in 2% KOH dissolved in 70% ethanol; wash in several changes of 70% ethanol.
2. Place ovules in mixture of 85% lactic acid and 70% ethanol (1:1) for 24 hours.
3. Dehydrate to absolute ethanol, then place specimens in 4½ fluid, to which 100 mg iodine is added per 9 g clearing fluid.

(Herr, 1971, 1974). Special depression slides are made as follows: 2 cover glasses (12 to 18 mm^2; numbers 0 to 2) are cemented to a slide 1 cm apart. Place clearing fluid with specimen in the gap or trough between the stacks. Place a third cover to bridge the gap. The supported cover glass allows a mountant thickness that eliminates excess pressure on embryos and other delicate materials. Using several cover glasses, the support stacks can be built up to any required thickness.

Phytolith Extraction

After comparing various techniques, Parr et al. (2001) and Parr (2002) describe a method of closed microwave digestion for isolating phytoliths from herbarium specimens. Phytoliths are usually non-crystalline, siliceous plant remains, or sometimes calcium oxalates, which are often of special interest to paleoecologists and archeologists. Numerous families of angiosperms have genus-specific forms of phytoliths. The authors achieved excellent results concentrating phytoliths from dried specimens. Material was treated with a concentrated acid mixture placed in pressure vessels in a pressurized microwave oven.

Use of this specialized equipment and the authors' lengthy protocol is beyond the scope of this guide. Refer to their reports for further instructions. See also Piperno's texts (1988, 2006). For details on using flotation and other processing of vegetation samples in archeological research, see the text by Pearsall (1989).

CHAPTER 4

Epidermis

A number of nondestructive replica techniques have been devised for examining leaf surfaces. Often these are sufficient for detecting epidermal cell patterns, making stomatal counts and aperture measurements, or characterizing morphology. In many taxa, however, this goal is frustrated when rough cuticles or dense populations of trichomes prevent a clear observation of underlying cells. Techniques for removal of cuticle or epidermis, and for leaf clearing as it pertains to stomatal visibility, are also included in this chapter. With these techniques, especially, careful monitoring of test samples is required to assess timing, solution concentrations, and heat.

Examining Epidermis and Stomata

Negative/Positive Replicas

The negative/positive replica system is good for sampling living leaves at different times of day, or for other experimental conditions, so that stomata may be scored as open, partially open, or closed. This type of replica is excellent where the goal is to see details of the cuticular surface in addition to epidermal cell types and outlines.

In the formulations that follow, all use volatile solvents. Many are likely unsafe to breathe, or should not be inhaled over a period of time. Use a fume hood or a chemical filter face mask, or at least work in a well-ventilated area. Proportions will vary depending on the nature of the cuticle being assessed.

Miller and Ashby (1968) used Zelitch's (1963) technique. Supplies are liquid silicon rubber, RTV11 (General Electric Silicone Products, Waterford, NY), and a catalyst, Nuocure 28 (Tenneco Chemicals, Inc.,

Lake Forest, IL). The time required to set is controlled by the number of drops of catalyst added to liquid rubber. Mix quickly and apply the mixture to the leaf surface. Setting should take seconds, after which the negative mold is peeled from the leaf. The replica surface is painted with nail polish and allowed to harden. The positive polish cast is placed on a slide. The cover glass is applied dry and held down with tape at the edges. This technique can be quickly done as a classroom exercise as well.

Martin et al. (1991) made dental latex molds resulting in negative replicas of the leaf surface. They used agarose to make positive replicas, which were then used to look at lens properties of epidermal cells and how they focused light on photosynthetic layers. The authors measured focal lengths and intensifications, which showed the light concentrated on the underlying photosynthetic layers to be 1.5 or more times greater than ambient light. Other anatomical details were also recoverable.

Weyers and Johansen (1985) reviewed the history of stomatal replica techniques. They used Xantopren Plus (Heraeus Dental, Hanau, Germany), a thin-flowing silicone-based precision dental material with an elastomer hardener. The base is a dimethylsiloxane, plus fillers, which is cross-linked by mixing an alkyl silicate and an organo-tin activator. The reaction releases a small amount of ethanol. See also Weyers and Travis (1981) and Bongers and Popma (1990).

1. Place ingredients on a glass slide and mix with spatula for 15 seconds.
2. Apply to abaxial leaf surface with minimal pressure. Hardening took 2–3 minutes.
3. Remove material from leaf surface.
4. Use nail polish to create positive impression. Let it dry for 1 hour, then remove from silicone negative.
5. Observe positive with light microscopy.

NOTE: The authors' goal was to determine aperture size, and the procedure is not necessary for stomatal counts alone. They experimentally altered the amount of hardener, which relates to the length of time the mixture remains at low viscosity before hardening. The goal was 2.5 minutes at 23°C. The highest efficiency for base to hardener was 20:1 to 6:1.

Groot (1969) used silicone rubber plastic, RTV11. Mix with a few drops of catalyst and spread over the leaf surface with a wooden spatula. After 2 minutes of hardening, remove the plastic with forceps and wash it in distilled water with dilute detergent. Blot the replica with filter paper (which has no loose fluff) and desiccate. When thoroughly dry, coat with clear nail polish in a dust-free environment. After 30 or more minutes, strip off this film, the positive replica, and mount dry on a slide. Cover it and seal the cover glass. See also Sampson (1961).

Procedure to Experiment with Negative/Positive Replicas

1. Supplies: (a) A negative replica material such as GE Clear Silicone II (Momentive, Columbus, OH), a 100% silicone sealant containing methoxypolydimethylsiloxane and four other silica-based ingredients; (b) Clear nail polish; and (c) Regular nail polish remover, containing acetone.
2. Select leaf. If wet, blot to dryness.
3. Apply silicone to leaf surface. After 50 minutes, silicone peels away from the leaf easily but may be left for 24 hours for proper curing.
4. Moisten silicone replica surface with 50% ethanol, or polish remover with cotton-tipped stick. Blot excess, then apply nail polish. Wait 15 minutes. If surface of silicone was too wet, nail polish will stretch if removed directly.
5. Cover the dry nail polish replica with a piece of transparent package tape and remove from the silicone. This supports the peel plastic and prevents stretching. Mount this unit onto a slide with replica side up. Observe dry. If desired, use resin mountant to tack down the corners of the cover glass.

Negative Replicas

Negative replicas by themselves are quite adequate if the goal is to count stomata and detect epidermal cell shape from leaves and other smooth plant surfaces. Techniques like the following are widely used.

Brewer (1992) suggested the following method, which works best with leaves that have easily viewable stomata.

Procedure

1. The organ to be sampled may be dried, restored, fresh, or pickled. If wet, blot dry the area to be sampled. If fresh leaves are used they must not be obviously wet.
2. Apply clear nail polish to the surface, and wait until dry (approximately 15 minutes). If left too long, it may be difficult to remove.
3. Remove the piece of polish. Occasionally this may be done by hand or by the edge using a fine forceps. The easiest removal technique is by using cellophane, transparent adhesive tape, or transparent package tape. Fold a piece of tape to make a sticky-free handle, with an exposed sticky tab on one end. Place the sticky tab on the edge of the dried polish and gently pull. The polish should readily separate from the leaf. One can also place the clear tape entirely over the dried polish.
4. Place the cast upside down on a glass slide. Place a dry cover glass without mountant. To make the mount permanent, place a tiny drop of nail polish at each of the four corners of the cover glass, just enough to hold the cover down but not enough to bond with the sample itself.

NOTES:

1. Placing the cast on the microscope stage, the stomata will be visible. For better contrast, you may have to lower the substage to produce some diffraction (and destroy Köhler illumination), or use oblique illumination (see chapter 8).
2. Beyond stomatal counts, it is sometimes possible to see clearly the outline of all epidermal cells. In other cases, it is impossible to detect the boundaries of epidermal cells, neighboring or subsidiary cells with any clarity. Thick cuticles often obscure cell edges. For good views of the cell borders, you may have to make paradermal sections or clearings (see chapter 1).
3. Some leaves are so obscured with hairs that good peels cannot be made directly. Often a new razor blade can be used first to gently shave or scrape the leaf surface in the area to be sampled. The resulting peel may not be pretty, but it will usually be informative.

Marx and Sachs (1977) studied mature leaves from nail polish replicas of leaf surfaces. The authors studied stomatal development in enlarging young leaves beginning at 1.5 mm long. They cleared and observed them in lactic acid without heat. The stomata were found to be nonrandomly distributed and to have a greater frequency on the abaxial surface by 3:1.

Runions (1998) used Krylon Ignition Sealant (Krylon Products Group, Cleveland, OH), which is sold as waterproofing for engine wiring. Spray the leaf surface, let it sit for 5 minutes; peel off the coating with clear tape. Place the taped replica on a glass slide and examine under phase contrast microscopy.

NOTE: See chapter 7 for alternative light microscopy (LM) techniques.

(Cheng et al., 2003). The authors used nail enamel to study maize leaf and stem surfaces. Take the peeled-off replica and float it in a drop of water on a slide, replica surface facing away from slide. Heat gently to flatten replica, draw off water, and let replica dry against the slide. Their replicas were observed microscopically with an "edge dynamic oblique illumination condenser" on an Olympus BX51 microscope. See chapter 7 for more on diffracted light contrast.

(Gloser, 1967). Methyl methacrylate is mixed to 5% strength in chloroform. The solution is painted on a leaf surface. When dry, the cast is removed from the leaf with transparent adhesive tape and examined microscopically. The method was used by the authors to measure stomatal apertures. Replicas will only work with fairly smooth cuticular surfaces.

Sinclair and Dunn (1961) used Archer's formula, a herbarium mounting medium that is no longer used because of toxic fumes. The authors adapted it for making leaf surface impressions. The formula, seen variously in the literature, is given here as published by Rollins (1955): 903 ml toluene, 184 ml methanol, 50 g Dow methyl styrene, e.g. Dow Resin 276 (Dow Chemical Co.), and 250 g ethyl cellulose, e.g. Dow Ethocel (Dow Chemical Co.).

Mix the two solvents, pour them into a beaker holding the resin, and stir until uniform. Add the Ethocel and stir until partially dissolved. Wait 24 hours before use. A separate mixture of the toluene and methanol

(4:1) can be used as a solvent for the resinous mixture. By experiment, use the 4:1 solvent to thin the resinous mixture to make leaf-surface impressions as required. Use a squeeze bottle or metal oil can to dispense Archer's medium.

CAUTION: Don't breathe toluene. Work only in a well-ventilated place if you choose this over nail polish. Not yet tried: in place of toluene, substitute the aliphatic limonene, e.g. Citra Solv (Citra Solv, LLC, Danbury, CT) or Histo-Clear.

Beerling and Chaloner (1992a) used cellulose acetate impressions to assess stomatal density. At an art supply store, obtain thin cellulose acetate sheets (ca. 0.003 ml) and cut into small squares of useful size. Place a few drops of acetone on the film, let stand a few seconds, and press the softened plastic carefully against the leaf surface. In 3–5 minutes the dried plastic film can be peeled from the leaf and viewed microscopically. (This is similar to the technique used to lift cellular detail from coal ball surfaces.)

Beerling et al. (1992b), studying the stomatal density of *Salix herbacea*, added a modification to the above. They flooded the leaf surface with acetone, then applied thin cellulose acetate film. Compress between flat surfaces for 10–15 minutes until acetone evaporates. The replica is mounted dry and photographed using transmitted light at 40X. Negatives are projected from 2×2s and measured on a projection screen.

Gitz and Baker (2009) developed a method for making stomatal impressions directly onto archivable plastic slides. The authors reviewed the history of a century of various impression techniques before suggesting the following. Four different plastic sheets were obtained, of which the best results were obtained using cellulose acetate (CA), available in a thickness of about 1/6″, and cellulose acetate butyrate (CAB), available in thicknesses up to 3/16″. The sheets were cut into 50×75 mm rectangular slides. A solvent, methylene chloride or acetone, was used to soften the plastic surface and a leaf pressed into it. The sheets can be cut to any size to accommodate whole leaves if desired.

CA yielded very detailed replicas but tended to warp upon drying. Warping is minimized when using very small leaves and when minimal amounts of solvent are used. CAB, using the same solvents, worked as well as CA and tended to warp less. The authors found that CAB is also cheaper than the thinner CA. Other plastics tested (with their appropriate solvents) were polymethyl methacrylate and polyvinylchloride. They did not produce the detail found in the ones recommended above.

Glue Impression Methods

White Glue

(J. Varner, personal communication). Obtain a white glue such as Sobo Premium Craft & Fabric Glue (Delta Creative, Inc., City of Industry, CA) or Aleene's fabric glue (iLoveToCreate, Fresno, CA). Spread it on one or more slides and allow to dry. A glass rod may be used to smooth the glue. Slides may be stored in a box until needed. To make a leaf impression, heat a dried glue slide on a warming tray. While warm, press a leaf onto the glue surface and remove. Allow to cool and examine slide under a microscope. Fabric glues remain more plastic than other white glues and become impressionable when heated. Other white or yellow glues might be tried.

Superglue

Judy Croxdale (personal communication) and Nels Lersten (personal communication) have noted that superglue (assuming cyanoacrylate) makes good leaf-surface impressions. Place superglue on a slide, press a leaf into it, and wait until it dries. Peel the leaf away from the slide.

Valkama, Koricheva, et al. (2005) used glue impressions to study the density of glandular trichomes of *Betula* leaves. A drop of superglue (cyanoacrylate) was placed on a slide. The leaf was pressed against the glue and then carefully removed after 1 minute. Trichome densities were counted from three randomly selected fields, 1.3×1.3 mm^2.

Valkama, Salminen, et al. (2003) placed a leaf surface onto a slide in a drop of superglue and removed the leaf after 1 minute. The resulting replica is clear and densities of trichomes can be counted. The authors have three methods pages. Birches have glandular trichomes on the leaf surface that are ecologically important. They contain flavonoids that are useful as UV radiation filters. Their study included transmission electron microscopy of surface glands. They counted three frames (0.65 mm^2) using "systematic uniform random sampling" (Kubinova, 1994). See chapter 8 for more on sampling techniques.

Epidermal Cells and Stomatal Counts

(Karabourniotis et al., 2001). When a fluorescence microscope is available, the surface of the leaf can be observed with little treatment as described below. The microscope is fitted with a G-365 excitation filter and an FT-395 chromatic beam splitter. The authors used a Zeiss Axiolab fluorescence microscope. Several species of leaves were handled one of two ways: (a) Some species can be viewed directly using the microscope as set up above. Leaves of *Olea europaea* were dehaired first using self-adhesive tape; or (b) Immerse the leaf in 10% KOH for 2 minutes followed by a distilled water wash. Then stomata may be easily viewed using the microscope.

NOTES:

1. The first method works in some cases, but some leaves must be carefully shaved or scraped if tape is found to disrupt the cuticle or epidermal surface.
2. Light-emitting diode (LED) illuminators are becoming more common for fluorescence work. See also chapter 8 for more on this topic.

The authors then described in detail the use of digital capture and image analysis, for which the article should be consulted. Using the above technique, they found that stomatal guard cells are often surmounted by cuticular ledges that emit blue fluorescence.

In the species tested, the induction of blue fluorescence was found to be caused by wax-bound phenolic compounds, rather than lignin. Lignin could be the fluorescing compound in conifer leaves. The authors review other methods of preparation for observing stomata.

Keating (unpublished) observed epidermal features from cleared and bleached leaf fragments that were mounted using calcium chloride. Abaxial or adaxial epidermal features such as stomata are easily characterized, photographed, and counted. This clearing technique works well for leaves that resist all known replica techniques. Abaxial surfaces of *Salix* leaves are frequently covered with conicoids, clusters of cuticular extrusions that obscure the shape and position of epidermal cells and stomata. Also, they act as reinforcing rods that hold fast to any known replica coating.

NOTE: Refer to chapter 3 for details. Also see related notes in chapter 8 for counting stomata from such preparations.

Phillips (2004) edited a column that collected comments from various researchers regarding leaf-surface observations:

Waxy cuticle removal. Use acetone to dehydrate, rather than ethanol. This leaves the surface clean, and cellular features stand out clearly. (Randy Tindall)

Trichome removal. This calls for a steady hand: Holding a razor blade at a slight angle, pull it across the leaf surface in the direction of the tilt. Avoid touching the epidermis. Or try holding the blade nearly horizontal and scraping across, again not touching the leaf. (Phil Oshel)

Surface replicas. Coat the leaf surface with nail polish, or a mounting medium such as SHUR/Mount (General Data Healthcare, Inc., Durham, NC), and strip off when dry. This usually removes trichomes and other surface debris, without damaging the leaf surface. (Ian Hallett)

NOTE: Hallet's method is worth a try, but trichomes of some species won't be pulled off by this method.

Dewaxing Leaf Surfaces

In many taxa, thick or waxy leaf cuticles obscure epidermal cell size and shape as well as stomatal density and typology. One approach involves peeling off the epidermis and looking at it from the "inside" (see Olowokudejo, 1990, and below), but this is not always easy to accomplish. Because some leaf surfaces may contain non-waxy ingredients such as lignin, you may need to try more than one of the suggestions below.

In *Microscopy Today*, Nov 2007: 60, Netnotes column: "Sample preparation-dewaxing leaf surface for SEM," various authors addressed the question of preparing leaf surfaces for SEM investigations. The suggested dewaxing procedures may be suitable for LM as well.

Suggested Dewaxing Procedures

1. Soaking leaf in xylene may not work, at least on papillose leaves. (Tina Carvalho)
2. Sluice the leaf surface with chloroform. (Sally Stowe)
3. Dehydrate leaf in an acetone series. (Randy Tindall)
4. After post-fixation water rinses, rinse leaf in acidified 2, 2-dimethoxypropane (DMP). The acidified DMP reacts with water (endothermically) to produce equal parts ethanol and acetone. This dehydrates the leaf and, as a side effect, dewaxes it. The acidified DMP is made by adding 1 drop of concentrated HCl (i.e., 0.05 ml) to 100 ml DMP. (Gilbert Ahlstrand)

Baker (1970) removed surface wax by washing with chloroform. Other miscellaneous cuticle dewaxing suggestions include (a) Dewax with cotton balls dipped in chloroform; (b) Sprays of pure seed oils will dewax leaf surfaces within 1–2 hours; (c) Paraffin oil, or sunflower oil; (d) WD-40 spray lubricant (WD-40 Company, San Diego, CA). This venerable product is 50% Stoddard solvent, which is a paraffin-derived white spirit. It also contains 25% liquid petroleum gas propellant (or CO_2), 15% mineral oil, and 10% inert ingredients; (e) Older dry-cleaning formulas that mixed aliphatic and alicyclic hydrocarbons; or (f) Paint brush cleaners or varnish thinner.

Cuticle or Epidermis Separation

Zinc Chloride and HCl Cuticle Isolation

Miller (1982), in a study of cuticular pores, prepared isolated cuticular membranes from apple fruits by taking paradermal pieces from the apple fruit wall. The pieces were aspirated in a solution of 3 g of zinc chloride in 5 ml of concentrated HCl. Just 10 ml of solution was used to process twelve 1-cm^2 pieces, with complete changes every 24 hours, until unbroken cuticles were obtained. After thorough rinsing in warm water to remove debris, the cuticles were air-dried, then stored in 70% ethanol. Later, after dehydrating the samples to 95% ethanol, cuticular waxes were removed using chloroform or xylene. Whole isolated cuticles were placed on a slide in 99% glycerol and cover-glass mounted.

Foster's Acid-Alcohol Cuticle Removal

Sinclair and Dunn (1961) examined cuticles from the inside. A leaf was soaked in water overnight, or boiled for 15 minutes. It was then treated in Foster's (1949) acid-alcohol macerating fluid, which is 70% ethanol and concentrated HCl (3:1). The authors removed the air with an aspirator. The leaf must be removed while still intact. If left too long, it will macerate. Blot the leaf surface dry and coat the surface in liquid plastic. Allow plastic to dry. The leaf mesophyll is then scraped from the epidermis, which is supported by the adherent plastic. It can be further soaked, stained, and mounted in resin.

Bleach-Treated Epidermal Layers

Mbagwu and Edeoga (2005) studied epidermal features of *Vigna* species. They immersed fresh leaves in 70% ethanol in test tubes. The tubes were placed in a boiling water bath for 4–5 minutes to help remove chlorophyll. Epidermal layers were peeled off and cleared in laundry bleach (sodium hypochlorite) for 2 minutes. Alternatively, the part to be examined was placed on glazed tile and flooded with the bleach for 2 minutes. Peels were then stained in 1% ethanolic safranin for 1 minute and mounted in aqueous glycerol solution. Stomatal indices and frequencies were calculated from photomicrographs.

Ethanol, Hydrogen Peroxide, and Ammonia

Christophel et al. (1996) studied Lauraceae cuticles prepared as follows.

Procedure

1. Place small pieces of leaf in test tube in 70–95% ethanol for 18 hours. (In this range, ethanol concentration is not significant.)

2. Remove ethanol by decanting or pipetting. Add 10 drops of 40% H_2O_2 and 5 drops of 90% ethanol. Simmer in boiling water bath 2–48 hours. Heating is complete when sample has turned light yellow to white and the cut edges of the cuticle begin to peel back like book covers.
3. Place samples in a watch glass or Petri dish. The upper and lower cuticles are pried apart and the internal cellular material can be brushed away with a small artist's brush. Or, before isolating and cleaning pieces, return them to 70–90% ethanol for 12 hours, which may render the cuticles more easily cleaned.
4. Rinse the cleaned cuticles in 2% ammonia to adjust the pH. They may be left unstained, or may be stained in 0.1% crystal violet for 20–120 seconds, until the cuticle appears uniformly light purple. If overstained, destain in 90–100% ethanol.
5. Mount in phenol glycerin jelly. If necessary, slides can be cleaned with a razor blade and the cover glass ringed with paraffin or nail polish for more permanence. These authors found ringing unnecessary.

Also using lauraceous leaves, Juan Carlos Penagos (personal communication) modified the above procedure as follows: (a) Use 3 parts of 3% aqueous hydrogen peroxide to 1 part 70% ethanol. This concentration of hydrogen peroxide is an over-the-counter drug store product; and (b) Heat specimen pieces in boiling water bath until epidermis separates. Specimens can be observed without staining.

NOTES:

1. Safranin staining also improves clarity of detail on some cuticles, especially when used with a green filter.
2. I find that unstained leaf surfaces or cuticles yield full detail using oblique illumination or a carefully lowered substage condenser.
3. Crystal violet may prove more photogenic without filters compared with Safranin O. Surfaces photograph well at 10X to 40X magnification.
4. This technique would work best with coriaceous leaves or chartaceous leaves with sturdy cuticles. It did not work for me when processing leaves with thin cuticles, which tended to disintegrate.

Chromium Trioxide Cuticle Separation

Alvin and Boulter (1974) suggested a maceration procedure that isolates the leaf cuticle for observations using either LM or SEM. After reviewing several techniques, they reported greatest success with their chromium trioxide technique. It yields what the authors call a "steady state" result.

To illustrate this, they note that Schultze's technique, which uses concentrated nitric acid and potassium chlorate, followed by ammonium hydroxide, can produce very effective cuticle isolation. Its problem is that immersion in these fluids must be timed exactly for your material. If left too long, features of the cuticle itself will dissolve. Alvin and Boulter's chromium trioxide technique stops acting when the cuticle is completely cleaned of cellulosic material, hence their term "steady state." This ensures comparable results over a range of timing and species variations.

Using taxodiaceous species, they experimented with chromium trioxide (a.k.a. chromic acid) at a variety of concentrations (5, 10, 20%), temperatures (20°, 40°C), and times (1, 6, 24 hours). The goal was to completely expose the granularity of the internal cuticle surface and to expose guard cells. Best results (steady state, complete cleaning of the surface) were obtained as follows: (a) 10% chromic acid at 40°C for 6 to 24 hours; or (b) 20% chromic acid at 20°C for 24 hours; or (c) 20% acid at 40°C for 6 to 24 hours.

Miller's (1986) later study of cuticular pores in leaf material noted that the zinc chloride/HCl fluid was not effective in their gymnosperm material. For these specimens, leaf pieces were aspirated in 10% chromium trioxide (20% in the cases of *Metasequoia* and *Cupressocyparis*), and placed in a 40°C oven for 12–24 hours.

Isolated cuticles were rinsed several times in distilled water and stored in 95% ethanol. To dewax, the cuticles were transferred to absolute ethanol (three 10-minute changes) and treated in xylene or chloroform. After air-drying, isolated, dewaxed cuticles were dry-mounted on slides where cover glasses were weighted until held in place by completely dried nail polish.

NOTE: Experiment with your material. For example, it may take only a few hours in 20% acid at 20°C to separate cuticles from thin leaves. Also it is necessary to dehydrate the specimens in an ethanol series before handling, as the cuticles are mushy-soft after the acid treatment. Use a pipette to transfer fluids to and from the specimen vial; don't attempt to transfer the specimen itself to a different solution. I've found this sequence of ethanol solutions to work well: 30, 50, 70, 95, 100%. In the last two solutions, the specimens can be lifted safely with a spatula and placed on a slide. See additional notes below for steps leading to SEM observations.

Epidermis Separation by Jeffrey's Solution

Olowokudejo (1990) used Jeffrey's solution to prepare leaves so that the stomatal index (SI), among other observations, could be derived.

Procedure

1. Remove samples (1 cm^2) from mid-point of leaf.
2. Soak in Jeffrey's solution: 10% nitric acid and 10% chromic acid (1:1), timing by experiment, then place in water bath for 15–60 minutes depending on sample thickness.
3. Abaxial and adaxial epidermal layers are teased from the mesophyll with fine forceps and dissecting needles, then transferred to 50% ethanol for 2 minutes to harden cells.
4. Stain in 1% Safranin O in 50% ethanol for 5 minutes. Each epidermis is dehydrated and mounted in resin.
5. Drawings can be made with a camera lucida. Stomata can be measured with a micrometer eyepiece.

Kong (2001) examined separated epidermal layers for viewing by LM. Leaf pieces were boiled, then macerated in Jeffrey's solution. The epidermal pieces were separated and then stained in 50% ethanolic safranin and mounted in glycerin jelly.

Scanning Electron Microscopy of Leaf Surfaces

Dried, mature leaf samples can be restored (see also chapter 3) or rehydrated in boiling water for 5–10 minutes, then fixed in FAA for 24 hours.

Wash samples in distilled water and dehydrate serially from 50% to 100% ethanol, air-dried and kept in a desiccator.

Following standard procedure, for good dimensional stability under the electron beam, specimens should be critical point dried after running through the ethanol series, then mounted on stubs, and gold sputter-coated.

Chemical Reaction Dehydration

Muller and Jacks (1975), Johnson et al. (1976), and Maser and Trimble (1977) have prepared dehydrated specimens with the use of 2, 2-dimethoxypropane (DMP). After fixation, specimens are given more than one rinse in water and then immersed in acidified DMP for times ranging from minutes to days. The DMP is prepared by adding one drop (0.05 ml) of hydrochloric acid per 100 ml of DMP. The reaction products when this mixture is introduced to a rinsed specimen are acetone and methanol. The reaction is endothermic, cooling the specimens. After dehydration, the specimens are suitable for any sort of further processing for TEM or SEM, such as critical point drying. Or, a translucent specimen can be directly resin-mounted for LM observation.

Alternatives to Critical Point Drying

(Keating). Recent experiments suggest that good SEM results are possible if a critical point drier is not available. Also, I've noted that a problem with critical point drying is that some leaf pieces, epidermal preparations, or cuticle pieces will curl up during the drying process. Use the technique below if curling is a problem.

Procedure for Keeping the Specimen Flat for Mounting on a Stub

1. Remove the leaf or cuticle piece from 95% or 100% ethanol, which can be handled directly with a small spatula at this stage.
2. Place the specimen on a small piece of plastic just larger than the specimen. A stiff plastic piece cut from a food container lid works well.
3. Wrap the specimen and supporting plastic in several thicknesses of lens paper strips and secure with tape or thread.

4. Place the packet overnight in dried silica gel. This will make the specimen anhydrous as well as preventing curling.
5. The specimen can then be handled with pointed forceps and placed on a stub to be gold coated.

NOTE: Check dimensions of the sample before and after the procedure, so you can correct your final measurements for shrinkage if necessary.

Brown (1993) suggested the following method for small flies, but it should be tried with botanical specimens. With specimens in liquid, dehydrate them to 100% ethanol. Change to hexamethyldisilazane (HMDS) and give the specimens 2 or 3 half-hour soaks. After the last soaking, pour specimens and HMDS into shallow dishes, watch glasses, or spot plates and let the HDMS evaporate under a hood.

Under the beam, the dehydrated specimens are said to resemble exactly those that have been critical point dried.

CHAPTER 5

Pollen and Ovules

The following methods are well adapted to making reference slides for systematic or developmental studies, and stress contemporary plants: herbarium specimens, liquid preserved, or freshly collected. See Hesse et al. (2009) for a superbly illustrated handbook of general pollen terminology. Traverse's (2008) text has an extensive treatment of methods, covering extant and fossil pollen, which also includes a discussion of other sources.

Sampling Contemporary Pollen

Atmospheric Samples

(Traverse, 2008). Microscopic pollen preparations can be made directly from an atmospheric sample. A microscope slide is coated or smeared with a translucent sticky substance such as Vaseline (Unilever, Rotterdam, The Netherlands), or glycerin, and exposed to the atmosphere. Pollen thus captured can be observed directly on the microscope stage. Without further treatment, many species can be identified as the dominant wind-pollinated species in the geographic area sampled. Such sampling is frequently accomplished by aerobiologists.

Pollen in Flowers

Characterization of pollen grains including accurate details of their wall and aperture structure frequently requires a cleaner sample. Oils, protoplasmic materials, and debris may obscure significant features.

The reason for further processing is that diagnostic features of exines are often obscured by oils, dust, and other debris. To make the surface features of the grain translucent and clearly visible, anthers or whole flowers can be boiled in 10% KOH. This hydrolyzes cellulose and removes contents and other organic materials. The (usually) resistant exine becomes lighter, or even colorless in some samples. Various stains can be applied to promote better microscopic visibility (see below).

When sampling plants in the field for pollen, remember the need to make pressed and dried voucher specimens. When stored in a herbarium, these specimens can be used many years later, if necessary, to check the identity of a specimen. See Traverse (2008) for extensive information on plant sampling for appropriate pollen stages for study.

KOH-Acetolysis

Because of its extensive history in pollen sample preparation, I offer these comments but no further description.

Traditionally, in many laboratories, pollen in reference-slide collections has been processed using KOH-acetolysis. The technique's special value with fossil materials is that it concentrates pollen grains dispersed in peat or organic sediment samples. The chemistry destroys all other organic materials in a sample, including pollen contents. This drastic treatment works because (most) pollen exines are among the hardest and most resistant materials in nature. For ease of comparison, reference slides from contemporary specimen vouchers have often been treated the same way. See Traverse (2008) for discussion.

CAUTION: It must be stated that acetolysis chemicals and fumes are corrosive and extremely toxic to breathe. The process must be carried out using a fume hood and with very careful handling. These safety techniques are detailed in Wodehouse (1935), Erdtman (1952), Brown (1960), Faegri and Iversen (1989), Traverse (2008), and Hesse et al. (2009).

For making pollen light microscopic reference slides of contemporary plants, KOH-acetolysis is not necessary. Using the preparation techniques below, samples from most species will yield good data on grain size, shape, aperture structure, and the exine sculpturing.

Pollen Staining

(Brown, 1960, p. 122). Most stains are made up by dissolving powdered dye in water, dilute ethanol, or more concentrated ethanol, depending on what the specimens are dissolved in. Dye concentrations of 0.1% to 1.0% are typical, as the concentrations matter little except for their effect on timing. Experiment with this on your material.

Brown (1960, p. 13ff) noted that pollen walls can be stained with fuchsin, crystal (gentian) violet, or methyl green after the Wodehouse process. If overstained, a rinse in solvent will usually destain the specimens to appropriate visibility. Try gentian violet, Safranin Y or Safranin O, methyl green, basic or acid fuchsin, or Bismarck brown.

NOTE: The use of dyes was very helpful for differentiating the exine in film-based photography. It may be less necessary for digital-image capture, as background vs. specimen contrast can be adjusted in Photoshop or other computer applications. (However, see section on photomicrography in chapter 8.)

Mounting Media

Media Miscible with Alcohol

Euparal is commercially available and has been a favorite for mounting small zoological specimens (Brown, 1960, p. 128). The recipe is said to consist of sandarac, eucalyptol, paraldehyde, camphor, and phenyl salicylate. Material must be dehydrated through 95% ethanol before mounting in Euparal. This saves the additional steps of moving through xylene or aliphatic hydrocarbons as required for mounting in Canada Balsam or commercial synthetic resins. With its RI of 1.481 being slightly lower than glass, Euparal will enhance edge contrast with translucent subjects.

Diaphane (Will Corporation, Rochester, NY) behaves similarly, but after 95% ethanol, the specimens should be rinsed in Cellosolve solvent before mounting.

Media Miscible with Water

Wet mounts can be made using glycerin or silicone oil. When using silicone oil, the pollen sample must be dehydrated through tertiary butyl alcohol before mounting. Such mounts will remain stable for long periods but must be stored flat and handled carefully.

Traverse (2008) noted that silicone oil has a favorable RI, but it never hardens. He recommends Dow Corning 200 Fluid (Dow Corning, Midland, MI), which is dimethylpolysiloxane, with a viscosity of 2000 centistokes. To mount pollen residues: wash with water; wash with drops of water and 95% ethanol; wash with 99% ethanol; stain if desired using safranin or fuchsin; wash in t-butyl alcohol. In a small vial, place material with silicone oil and 1 ml of tert-butanol. Allow to evaporate for 24 hours. Samples can be stored in these vials. Mount material in a small volume of silicone oil. An advantage of a non-hardening medium is that the cover glass, and therefore the grain, can be rolled around to present the various features.

Dahl (1952, p. 252) suggests mounting pollen in oil to observe the contracted air-dry state. Then, soften anthers in 10% KOH or NaOH. Transfer anthers to 85% lactic acid.

(Maurizio, 1953, p. 48; Brown, 1960, p. 121). Anthers or pollen from anthers are placed on a glass slide. Add drops of ether or chloroform to remove lipids. Evaporate or decant solvent; remove debris. Add glycerin jelly. Seal edges of cover glass with mounting resin or lacquer.

Glycerin Jelly

This semi-permanent mounting medium can be applied to pollen samples removed directly from water or alcohol. To make the slides permanent, the covers must be ringed with resin, nail polish, or paraffin. In hot climates it may not solidify well. While glycerin jelly's RI is favorable (1.43), it is not truly an archival medium.

Ogden et al. (1974) note that pollen may be stored in this medium in capped vials. Glycerin jelly may be pre-stained by adding basic fuchsin at 1 drop per 15–20 ml of glycerin jelly. A stock supply of glycerin jelly may be stored in a stoppered flask or capped vial and can be heated for use in a water bath. It should not be heated repeatedly. One procedure involves removing a chunk of solid jelly from stock as needed to make a slide that day. Place it on a slide and warm gently until melted, then stir in a pollen sample and cover. Or, if many reference slides are to be made, warm a vial of jelly in a water bath.

The authors recommend: gelatin 2 parts; water 12 parts; glycerin 11 parts; and 2% phenol (1 part to each 50 parts of glycerin jelly). Mix gelatin and warmed water; when dissolved, add glycerin and phenol, let stand overnight, strain with cheesecloth. They recommend using the same mountant and prep method for reference slides as for unknowns or specimens to be investigated.

Traverse's (2008) glycerin jelly mixture: gelatin 50 g; glycerin 150 g; phenol 7 g; and distilled water 175 cc. Dissolve phenol in warm glycerin and dissolve gelatin in warm water (add gradually while stirring); allow time to dissolve. Mix both liquids together and warm to no hotter than 80°C. Bottle in wide-mouth jars and store at room temperature.

Other Mixtures Described in the Traverse Volume

1. Kaiser's glycerin jelly: gelatin 40 g; water 210 ml; glycerin 250 ml; phenol 1 g. Soak gelatin in water 2 hours. Add glycerin and phenol. Heat 10–15 minutes, stir until uniform; cool. Refrigerate in storage; melt to use as needed.
2. A variant of above: gelatin 1 part; water 4 parts. Allow mixture to soak for 2 hours. Add glycerin 7 parts by weight; phenol crystals 1% of total mixture volume. Warm total mixture 15 minutes; filter through cotton filter, if needed.
3. Sass: gelatin 5 g; water 20 cc; glycerin 30 cc; phenol crystals 5 g, dissolved in 10 drops of water.
4. Gelatin 1 pkg. (ca. 7 g); hot water, enough to dissolve packet. Stir. While hot, add 50–100 ml glycerin. Stir until uniform.
5. Gelatin 10 g; water 60 ml. Soak 2 hours. Add glycerin 70 ml; phenol 1 g. Heat in water bath; mix; cool. Heat to use.

Sugar Syrup Mountant

(Brown, 1960, p. 135; cites various authors). Karo (Associated British Foods, London, UK) or other corn syrup, brown or clear. Dilute to ⅔ normal strength. The solution will harden on a slide overnight without a cover glass. Mountant can be softened by adding drops of water so that specimens can be reoriented. A cover glass can be placed over the fresh syrup to make the preparation permanent. Fourteen-year-old slides were in good condition, although some crystallization occurs around the edges. This can be fixed by adding water and gently warming the slide. Phenol or other toxicant may be added, especially in warm climates.

Alternatively: Karo syrup, water 1:1. Add 1–2 crystals (=0.2 g) thymol as preservative to each 80 ml total.

Quick Pollen Slides

Wash pollen in 50% isopropanol or ethanol (after fresh, pickled, or cleaned). Hydrate in 50% glycerol. Transfer to glycerin jelly (a piece of hardened jelly is melted on a glass slide) and add cover glass. For more permanence, use a heated probe to feed melted paraffin under cover glass edges.

Lactic Acid Based Recipes

(Dahl, 1952; Cranwell, 1953; Canright, 1953). This medium expands dried pollen from herbarium specimens to approximately fresh size. They do not last well enough to serve as reference slides (Brown, 1960, p. 132). An advantage of lactic acid is its RI, which, at 1.42, is close enough to that of glass to provide a clearer microscopic image. Use 85% aqueous lactic acid.

Amman's lactophenol. (Brown, 1960, p. 132). Phenol crystals 20 g; lactic acid 20 g; glycerin 40 g; and distilled water 20 ml. This is a good temporary medium into which stain can be incorporated. Slides must be stored flat.

Gum Arabic-Glycerin Medium

(Brown, 1960, p. 134). This medium will harden so that cover glass will not slip after 24 hours on a warming plate. In 3–4 days, pollen slides may be stored safely on edge. Slides 10 years old were in good condition. Mixture: Gum arabic lumps 50 g; water 100 ml.

Suspend lumps in gauze bag in water. After 3–4 hours, heat to boiling and allow gum arabic to drain from bag.

Mix 3 parts gum arabic solution to 1 part glycerin; add phenol crystals to 1%. Glycerin/gum arabic ratio may be altered to make the solution more fluid or viscous. Powdered gum arabic as an ingredient does not work as well; slides dry out in 1–2 years.

Hoyer's Mountant

(For recipe, see chapter 7.) This mountant destains pollen due to the presence of chloral hydrate. Brown (1960, p. 123) notes that apparently fungal mycelia will retain stain in Hoyer's.

Resin-Mounting

Currently, the mounting resins with the best reputation are synthetic resins such as Permount (Electron Microscopy Sciences, USA-Hatfield, PA) or Histomount (National Diagnostics). The traditionally used natural products such as damar gum or Canada Balsam are still quite useful, although they tend to yellow with age.

The resins are dissolved in hydrocarbon-based ring-solvents such as xylene, toluene, or aliphatic solvents such as limonene (Citra Solv or Histo-Clear). As a group, such resins yield the most permanent, archival preparations. Because they are immiscible with water or alcohol, the pollen samples must be dehydrated through a series of changes through 95% ethanol, absolute ethanol, then into the hydrocarbon solvent. For many purposes this extra work may be unnecessary in light of the aqueous-based techniques emphasized in this chapter.

(Ikuse, 1956, p. 2). Dissect pollen from the anther into a drop of 95% ethanol, heat gently to evaporate the ethanol. Add either gentian violet or fuchsin in 95% ethanol. Wash and dehydrate with absolute ethanol. Ikuse recommends transferring pollen to xylene, then mounting in resin.

Mountant-Stain Combinations

(Brown, 1960, p. 123). Safranin can be incorporated into glycerin jelly at 1% concentration. The walls of pollen mounted in this mixture will absorb the dye from the surrounding mountant within 1–3 days. Pollen usually becomes deep red, fern spores become lighter red, and sphagnum spores show no color change.

Calberla's Solution

Ogden, Raynor, et al. (1974), and R. R. Clinebell II (personal communication) modified Calberla's Solution as follows: (a) Stock solution A (can be stored in light): glycerin 5 ml; 95% ethanol 10 ml; and distilled water 15 ml. (b) Stock solution B, a saturated solution of aqueous basic fuchsin (store in the dark): add a scoop of basic fuchsin crystals to 20 ml distilled water. Stir to make a "blood red solution." Filter and discard solids. (c) Working solution (lasts indefinitely if stored in the dark): In a small dropper bottle, place several milliliters of solution A. Add several drops of solution B to make a "light rose" solution. Add more or less of solution B to achieve a pink exine, not dark red.

Procedure

1. Place pollen sample from insect or anther on slide.
2. Add a drop of working solution, stir with needle to mix well with sample.
3. Let sit for about an hour, then place a cover glass. Pollen will be fuchsia-colored; other debris will not take up dye.
4. The pollen exine should be stained pink, not dark red. Adjust the amount added of solution B to achieve this.
5. To harden the mount, mix a few drops of glycerin or melted glycerin jelly to the working solution. Store in an airtight jar. Refrigerate the glycerin jelly if storage is required for greater than several months.

NOTES:

1. The working dilution should sit as a meniscus on the slide and not run over the surface. If too runny, try one of the following: (a) Degrease the slide by washing first with soapy water or soaking in 50–95% ethanol cleaning solution; wipe dry; (b) Add one or more drops of glycerin to the working solution to add viscosity; or (c) Gently heat gelatin in equal volume of water; stir; add drops to working solution until it becomes adequately viscous.
2. If pollen grains are especially numerous on the slide, the dye may become too diluted to stain grains adequately. Fix this by: (a) Making

mounts with smaller quantities of pollen; or (b) Adding more of solution B to the working solution to make it darker.

3. Slides must be stored flat, but if sufficient glycerin jelly is present in a Calberla's solution mount, the slide should handle easily when using immersion oil under a microscope objective.

Acetocarmine and Glycerin

Gerlach (1969) suggests the following for observing generative and sperm nuclei. Place fresh pollen grains in a drop of acetocarmine; warm the slide a few seconds. Add glycerin and cover, then observe with LM as usual.

Basic Fuchsin and Glycerin Jelly

G. Van Brunt (personal communication) offers the following method for preparing reference slides.

Procedure

1. Place a clean glass slide on a warming plate heated to about 60°C. Place a drop of melted glycerin jelly on the glass slide and spread the jelly around in a smooth circle with an artist's brush. Remove the slide from the warming plate and let the glycerin jelly harden. (The circle of hardened glycerin jelly should have a diameter of about 18 mm if a 22-mm cover glass will be used later.) Glycerin jelly slides can be stored in the refrigerator in a slide box for at least a year.
2. Using a small brush, collect pollen and place sample on the hardened glycerin jelly surface. Most pollen will adhere to this surface.
3. Using Coplin jars for the solutions mentioned below, transfer the slides as follows: 100% isopropanol (to remove pollenkitt, slides can be stored here for weeks). Transfer slides to 75% isopropanol, then stain with basic fuchsin in 50% isopropanol.
4. Prepare a 10% stock solution of a saturated basic fuchsin solution.
5. Prepare the working solution by placing 6–12 drops of the stock solution in a Coplin jar of 50% isopropanol. Use test slides to ascertain concentration and times.

6. Rinse in 50% isopropanol or water, 10 seconds, if necessary.
7. Dry the slides. Add 1 or more drops of concentrated fructose (100 g fructose in 40 ml distilled water) solution, and then place a 22-mm cover glass. The fructose mountant should reach the edge of the cover glass. This concentration of fructose has an RI of 1.47 and has the advantage over glucose and sucrose in that it tends not to crystallize.
8. Slide may be made permanent as follows: Add 3 coats of shellac around the edge of the cover glass. Follow this with 2 coats of polyurethane. Slides thus prepared have lasted 4 years, so far, with no deterioration. Oil can easily be removed from slides with Sparkle glass cleaner (A. J. Funk & Company, Elgin, IL).

NOTE: If using tools to prepare multiple pollen preparations, it is especially important to clean all tools thoroughly after each sample to avoid contamination.

Size Variation

A relevant question is: What happens to pollen-grain size when using various mounting media? This is worth some experimentation if small differences are at issue in your research. In Canright's test (1953), he mounted dry pollen from herbarium specimens in lactic acid, which expands the pollen. Pollen grains thus mounted are only 1.6 μm larger than pollen prepared by Wodehouse's glycerin jelly process. Erdtman's grain measurements are 7.5 μm larger than pollen expanded by lactic acid.

Ovule and Small Tissue Clearing

The clearing and squash techniques below were a major contribution to the study of embryology by Herr (1971, 1972a, 1972b). For an example, see Smith (1973). Detailed reviews are found in Radford, Dickison, et al. (1974) and in Herr (1993).

Ovules or small ovaries (ovularies) are dissected and fixed for 24 hours and stored as needed, in 70% ethanol. For clearing, the specimens are placed in Herr's 4½ fluid for 2–24 hours.

The solution is prepared as follows: 85% lactic acid, chloral hydrate, phenol crystals, clove oil, and xylene (2:2:2:2:1), by weight. Try this treatment first before the modifications listed below. Rudall and Clark (1992) substituted the less toxic Histo-Clear (limonene) for xylene with identical results.

The 4½ treatment renders the ovules nearly transparent using bright field microscopy. Other types of optics may work better for some materials such as phase contrast, circular oblique lighting, or oblique lighting (see chapter 8). The shallow depth of field that this fluid allows makes possible a through-focus series to identify nuclei and stages of division. Ideally, Nomarski DIC or Hoffman modulation contrast optics set the standard, but the simpler optical methods named above can be made to perform well.

For clearing, specimens were placed in Radnoti planchets (no longer made), cell chamber slides, or well slides. After treatment, ovules or other small parts can be transferred by pipette to Raj slides made as follows: Two stacks of ca. 18-mm square cover glasses (No. 0–2) are cemented to a microscope slide about 1–1.5 cm apart. On a warming tray, allow the cemented stacks to harden. Usually 1–3 covers per stack are sufficient, depending on the size of the ovules. The specimens are placed, in their fluid, in the space between the stacks. A cover glass is placed, as a bridge, across the space.

If the space is the right thickness, the cover will not squash the ovules, but the cover can be tapped with a dissecting needle to reorient or apply gentle pressure on the ovules.

Clearing Modifications

Large, taxonomically related variations may be found in ovule size, shape, and transparency, so the original 4½ fluid may not be sufficient. Herr's summary of the techniques in the Radford et al. (1974) reference includes a number of clearing pretreatments and alternative clearing mixtures. Many of these are listed independently in Herr's references. See especially Herr (1982) for a review of these techniques. In the following modifications, the user may find improvements in contrast between nuclei and cytoplasm and cell layer delineation:

1. IKI-4½ (Herr, 1972a). To 9 g of the 4½ mixture, add 100 mg iodine and 500 mg of potassium iodide. This improves the phase contrast image as well as detects any accumulated starch.
2. PP-4½ (Herr, 1973a). To 1 g of the 4½ mixture, add 3 mg of potassium permanganate. Mix immediately before use.
3. BB-4½. (Herr, 1973b). Add 1 part benzyl benzoate to the standard mixture.
4. PPBB-4½. (Herr, 1973b). Just before immediate use, add 3 mg of potassium permanganate to 1 g of BB-4½.

Pretreatments. Before using the above methods, pretreat with any of the following: (a) Lactic acid; (b) Lactic acid/phenol/benzyl benzoate; or (c) Potassium hydroxide.

CHAPTER 6

Dyes and Tissue Detection

By traditional usage, dyes are termed stains when used in biological histology. Stains are identified below using nomenclature from Lillie et al. (1969) and Horobin and Kiernan (2002). They are traditionally used to provide "amplitude objects" that clarify tissue boundaries in microscopy and photomicrography. They are used much less frequently as histochemical tests. In any event, histochemical screening requires several forms of corroborating evidence (Jensen, 1962).

Dye Behavior

Dyes can be called *orthochromatic* when they react to a tissue with the color typical of the dye in solution. They are *metachromatic* when the perceived wavelength (transmission spectrum) of some of the dyed tissue is shifted to a longer wavelength. Such dyes are also called *polychromic*. For example, cresyl violet acetate (CVA) makes a purple solution in water or dilute ethanol. In the presence of calcium chloride solution, the dye becomes blue or green when binding to lignin, tan when binding to cellulose, and red when binding to pectic materials.

A Note Regarding the Procedures Below

Many authors who recommend aqueous dye solutions suggest adding various preservatives. The idea is to prevent mold and spoilage of these mixtures. Preservatives include phenol crystals, sodium salicylate, aspirin (sodium acetyl salicylate), and others.

While preservatives are useful, my tests suggest a simpler approach. I have found that most such dye powders recommended for "aqueous" solutions dissolve well in 15% ethanol. These mixtures remain stable for long periods, pipette well, and, due to their reduced surface tension, do a better job of penetrating tissues. While some dyes will precipitate after some months in solution, there is no mold spoilage. It is best to store the solutions in glass vials; polyethylene, and perhaps other plastics, will allow ethanol to evaporate through its porous wall.

Many of these dyes, referred to as *coal tar* or *aniline dyes*, are regarded as potentially carcinogenic by various authors (cf. Vierheilig et al., 1998). Handle the powders and solutions carefully and keep away from skin, especially from under fingernails.

Tissue Detection

Lignin

Foster (1949) recommended the use of a saturated solution of phloroglucinol in 18% hydrochloric acid. This dye will turn lignified materials (walls of fibers, tracheary tissue, and sclereids) bright red. Mount a section in this reagent and add a cover glass. Prepare the section away from the microscope since hydrochloric acid should not contact any part of the microscope.

(Breil, 1965). Using a razor blade, longitudinal and transverse sections (LS and TS, respectively) are easily cut from soft materials and are transferred to a drop of phloroglucinol on a slide. After 2 minutes, place a cover glass. The color change is immediate as lignified walls turn bright red.

NOTE: The formula for the above procedure is phloroglucinol 2 g per 100 ml of 25% hydrochloric acid solution. Other authors have suggested separate bottles of aqueous phloroglucinol and acid, where a drop of each is added to a slide. However, the acidified dye mixture is fairly stable. Herr (1992) made an aqueous solution of phloroglucinol and calcium chloride. He added a few drops of hydrochloric acid but considers any acid satisfactory. After years, this mixture still gave a good test for xylem, and Herr detected no deterioration of the solution with time. Herr noted that hydrochloric acid serves to chlorinate the lignin, as does the calcium chloride, which then develops the red reaction. Therefore, when using calcium chloride, other acids complete the reaction.

Alternative Formulas

1. Phloroglucinol 1% in 20% calcium chloride (20 parts); concentrated nitric acid (2 parts).
2. Phloroglucinol 1% in 20% calcium chloride (15 parts); lactic acid (5 parts). Mount the slide in straight calcium chloride, or in this solution.
3. Phloroglucinol 1% in 20% calcium chloride (25 ml); concentrated hydrochloric acid (4 ml). Sections can be transferred to straight calcium chloride.

NOTE: While acidified phloroglucinol solution may turn dark with time, it remains efficient.

Herr (1992) also found that Toluidine blue O (TBO) is just as good a lignin indicator. In calcium chloride it requires no added acid and gives identical (though blue) results, the color being removed from all other tissue.

Kutscha and Gray (1972) noted that Wiesner, in 1878, devised the application of phloroglucinol treated with strong HCl, such as 1 drop of 35% HCl on a slide with the section. Therefore, one sees reference to Wiesner reagent in contemporary literature. With strong acid, it is said that phloroglucinol detects lignin components as follows: red-purple for coniferyl aldehyde, blue-purple for synapyl aldehyde, and peach for syringaldehyde.

De Fossard (1969) discusses various chemical tests for lignin including the phloroglucinol/HCl and Mäule reactions. Distinctions among syringaldehyde, sinapaldehyde, and cinnamaldehyde are possible but are dependent on various fixation formulas. Kutscha and Gray (1972) describe this Mäule procedure as follows. This stains guaiacyl lignin dark brown and syringyl lignin pink-red.

Procedure

1. Immerse sections in 1% $KMnO_4$, 5 minutes.
2. Rinse in distilled water, 3 times.

3. Immerse in 3% HCl, 1 minute.
4. Rinse in distilled water, once.
5. Immerse in 25% NH_4OH, 1 minute.
6. Rinse in distilled water.

Toluidine Blue O (TBO) Dye

TBO is a cationic, metachromatic dye that binds with negatively charged groups. Gahan (1984) notes that the dye is orthochromatic in solution below pH 3 and metachromatic above pH 3. When applied to plant sections as a neutral aqueous solution, TBO will yield various hues as follows:

Color	*Produced in reaction with*
Pinkish purple	Carboxylated polysaccharides (e.g., pectic acids)
Purplish or greenish blue	Macromolecules with free phosphate groups (e.g., nucleic acids)
Green, greenish, or bright blue	Polyphenolics such as lignin, tannins (xylem walls: blue-green; phloem fibers: bright blue)
Not stained	Hydroxylated polysaccharides (e.g., starch, cellulose—some phloem elements)
Pinkish purple	Polyuronides such as primary walls of epidermis or parenchyma (sieve tubes and companion cells)

Evered (2009) and McAuliffe (2009) suggest adding 1% sodium tetraborate, commercially available as Borax (Rio Tinto Group, London, UK), to a TBO solution. The recommendation was specifically for use when staining tissues embedded in resins for TEM work.

NOTE: I have not yet tried this, but: in place of Borax, one might try substituting sodium hexaborate, readily available as Kodalk, or Kodak balanced alkali (Eastman Kodak Company).

Cellulose

Foster (1949) and Rawlins (1933) recommended a specific test for cellulose that uses potassium iodide and H_2SO_4. Mount section in I_2KI solution (iodine 1 g, KI 3 g, distilled water 300 ml). Add cover glass. An added drop of 65% H_2SO_4 will cause cellulose to turn blue.

Artschwager (1921) and Venning (1954) recommend zinc chlor-iodide metachromatic staining. This technique was recommended for fresh or frozen sections. Cellulose turns bright blue, with sieve tubes staining more intensely than surrounding parenchyma. Lignin, cutin, suberin, and immature xylem are yellow-brown.

Procedure

1. Prepare solution A: iodine 1 g; potassium iodide 1 g; water 100 ml.
2. Prepare solution B: zinc chloride 2 g; water 1 ml.
3. Stain sections in solution A for a few seconds.
4. Transfer to solution B. Stir material around gently until cellulose is bright blue.
5. After solution B, more of solution A can be added if necessary to intensify the blue reaction.

Schaede (1940) recommended the following solution: 15 g zinc chloride is dissolved in 10 ml of water. Add 0.15 g potassium iodide and 0.25 g iodine. The iodine quantity assures a concentrated solution. Allow the solution to develop an amber color, about 2 days at room temperature. With parenchyma membranes of pith and root tissues, it develops a pure dark-blue color in fresh cut sections.

Maácz and Vágás (1961) developed a triple stain method for plant sections having cellulosic and lignified cell walls. They hold that the combination makes it possible to detect fine changes as cell walls undergo lignification. It can be used after paraffin sections are brought down to water, or with sectioned material not originally embedded. Material is originally fixed as usual. The three stains are used in the order listed below.

Procedure

1. Astra blue 0.5 g, in 100 ml of 2% aqueous solution of tartaric acid: 1–5 minutes.
2. Distilled water rinse.
3. Auramine O saturated aqueous solution: 1–5 minutes.
4. Distilled water rinse. Sections have a green tint at this point.
5. Safranin O, 1% aqueous solution: 1–5 minutes.
6. Distilled water rinse.
7. Acetone rinse, 3 times, then phenol benzene 1:3, followed by resin-mounting.

RESULTS: Cellulosic walls are stained an intense blue. Tracheary tissue becomes red to orange to yellow, and the transition between lignified and blue cellulosic walls is very sharp. Cytoplasm is colorless while nuclei are bright red. The authors claim it is not possible to overstain even when stain periods are doubled.

NOTE: For the sake of safety, try a limonene solvent and limonene-based resin mountant, rather than benzene.

Sieve Plates

Sieve areas and sieve plates can be very difficult to detect, even in well-made sections. Foster (1949) recommended Aniline Blue WS as specific for callose on sieve plates. For a short time immerse sections in 0.1% aqueous aniline blue solution, rinse, transfer to a drop of water. Sieve plate callose turns blue. Foster noted A. S. Crafts' modification: Place section in IKI, rinse in water, stain in aniline blue 5 minutes, wash briefly again in IKI. Mount in water or glycerin.

Living Cell Primary Walls

Foster (1949) recommended neutral red for detecting living cell primary walls. Mount sections in 0.1% aqueous solution of the dye. This dye is also used as a general vital stain that detects intact vacuolar contents.

Cuticles, Suberin, and Exodermis

Foster (1949) recommended Sudan IV as a specific test for cuticle and suberized cell walls. Make a solution of 0.5 g Sudan IV to 100 ml of 80% ethanol. Place a drop on the sections; add a cover glass. Cuticle and waxy materials in walls are stained red.

Gahan (1984) noted that suberin also stains with Sudan Blue II, or also with gentian violet.

Seago et al. (1999) used Sudan VIIB, or "fat red." It is suberin-specific, giving a red reaction. The exodermis is a layer of hypoderm with Casparian strips oriented like endodermal bands. The hypodermis is the same layer without these thickenings.

Tannin Cells

(Keating). Add caffeine (trimethyl purine), in white powder form, to 1% strength in fixative or in 30% ethanol. Place hand-sections of a plant organ into a small staining dish or on the surface of a slide. Add a few drops of the caffeine solution. Any idioblastic tannin-containing cells will turn red. The detection may only last several days, but the effect is the same in an aqueous mount or in a calcium chloride mount.

See Iwanowska et al. (1994), who used caffeine in glutaraldehyde fixative for EM work.

Gahan (1984) noted that tannins turn blue-black in acidified ferric chloride solution. Use ferric chloride 1% in 0.1N HCl. Recommended time: 15 minutes.

Osmophore Staining in Orchid Flowers

Stern et al. (1986) explored the use of neutral red and Sudan Black B for detection of osmophores in orchids. These are surficial "glandular" areas, often on adaxial corolla lobes, that emit floral fragrances. Neutral red was used to detect osmophores in living orchid flower parts, and Sudan Black B was useful for sectioned, epoxy-embedded materials. In the orchids sampled, osmophore staining occurred intensely to vacuolar contents, especially of epidermal cells.

Neutral red is a dye used traditionally as a vital stain for vacuolar contents (cell sap). With orchids, some terpenoids are implicated as floral fragrances. The authors hypothesized that the amino group of the neutral red chromogen may form an electrostatic bond with the hydroxyl groups of some terpenoid fragrance compounds.

Sudan Black B is a widely used lipid stain. Some orchid osmophores stain positively with neutral red and negatively with Sudan Black B, or the opposite. As with many histochemical techniques, the chemical specificity of these dyes must be interpreted cautiously. In taxa where floral parts show a neutral red reaction, the extent and position of such a reaction may be systematically and functionally meaningful.

Procedure

1. Formula: neutral red 1 part; tap water 1000 parts.
2. Immerse whole living flowers in neutral red solution, 20 minutes for orchids; up to 10 hours in other flowers.
3. Rinse specimens in tap water. Vogel and Cocucci (1988) found that prolonged tap-water rinse tended to destain. Stern et al. (1986) did not observe this.
4. Compare stained flower with unstained flower of the same species to assess results.
5. Stained areas of specimens can be observed: (a) Using a hand lens or dissection microscope; (b) As a slide-mounted dissection or hand-section using a compound microscope; or (c) Dehydrated, epoxy resin embedded, and sectioned at 1 mm for microscopic analysis. If resin embedments are made, Sudan Black B preparations should also be made.

Calcium

There appear to be no simple techniques for calcium detection, especially calcium oxalate. Diagnosis can involve x-ray diffraction, polarization, SEM equipment, or more complex chemistry. Ilarslan et al. (1997) successfully used the Sigma Urinalysis Diagnostics Kit (Sigma-Aldrich Corporation, St. Louis, MO), procedure No. 591 protocol. See literature cited in Ilarslan et al. Yasue (1969) reviewed chemical methods that identify calcium oxalate in kidney tissue. There are several steps that involve, first, dissolving other calcium salts.

Acetic acid and hydrochloric acid will dissolve calcium carbonate and calcium phosphate. Then immerse material in 5% aqueous silver nitrate 10–20 minutes. Rinse in distilled water. Immerse 1 minute in saturated rubeanic acid in 70% ethanol, with added 2 drops of ammoniac water per 100 ml. Rinse in 50% ethanol and water. Calcium oxalate deposits will be dark brown or black.

Xylem Flow Detection

Silk, in an oral presentation not further referenced in the abstract (Baum, Silk, et al. 1996) found that eosin, a water-soluble fluorescent dye, is taken up by vascular bundles. The dye solution will diffuse out from protoxylem or metaxylem, whichever is functional. Metaxylem becomes functional only after leaf expansion is complete.

Arbuscular-Mycorrhizal Fungi

Vierheilig et al. (1998) reviewed the various dyes that have been used to detect AM fungi in the roots of land plants. Since some of these compounds could be considered hazardous, the authors successfully experimented with fountain pen or drafting inks, used in combination with white vinegar. The technique works to detect fungal infections equally well in leaves and in roots. Their comparisons of the ink/vinegar technique against the usual use of trypan blue were very favorable.

Procedure

1. Boil roots in 10% KOH to clear.
2. Rinse in changes of tap water.
3. Boil roots for 3 minutes in 5% ink in white vinegar, the latter being 5% acetic acid.
4. Rinse roots in tap water that has been acidified with a few drops of vinegar.

NOTE: Destaining times varied with the type of ink. Best photographic contrast was obtained with Sheaffer black ink (Sheaffer Pen Corporation, Shelton, CT). After destaining, the cleared roots remained a pale brown-red and the fungal structures remained black. Inks of various colors were tested from Waterman, Reynolds, Parker, Lamy, Pelikan, and Sheaffer. All were successful to some degree. After storage for 6 weeks in tap water, the specimens showed no deterioration in clarity.

Selected Dye Properties

I experimented with numerous dyes using calcium chloride as a mountant. The table on pages 130–131 shows the most promising results in terms of differentiation or polychromic properties. All dye samples were dissolved in 15% ethanol, at a strength that made them translucent in a 4-dram vial. They were applied to hand-sections of TS petiole or primary stem tissues, either fresh cut or fixed in FPA_{50}. Following the dye treatment, tissues were rinsed in 15% ethanol and mounted in calcium chloride. For tests, thin TS slices of celery petioles worked well, as did stem sections of *Epipremnum*.

Bismarck Brown Y

Venning (1954) pointed out that this dye differentially stains tissues or may attach to certain cell constituents, often removing the need to use counterstains. (If Venning means metachromasy or differential staining, I failed to note it.)

Blaydes (1939) suggested that keeping properties are improved by adding 5% phenol to a 1% aqueous solution of the dye. (Bismarck brown 1 g, phenol crystals 5 g, distilled water to 100 ml.) It has been used with hematoxylin as a counterstain in staining schedules for resin-mounted slides.

Chlorazol Black E

Darrow (1940) noted exceptionally sharp differentiation and some metachromasy with this dye. Stain sections in a 1% solution of the dye (in 50–70% ethanol). No mordant or differentiation is required. Results from three examples: (a) *Puccinia*: cell outlines jet black; cytoplasm and plastids grayish-green; nuclei dark green; hyphal infections yellowish green. (b) *Fern leaves*: cell walls black; nuclei green; cytoplasm light amber; plastids gray; suberized walls dark amber. (c) *Onion root*:

cell walls dark gray to black; cytoplasm grayish-green; nuclei yellowish-green; nucleoli yellow or amber.

Powdered Dye Stability

Titford (2002) noted that in a long-established institution, one may run across old bottles of dye powders. Encountering dusty bottles with faded labels may make one wonder if they could be useful in histology. Should the dye be discarded, or is it worth spending the time to make up a solution? With the costs of today's dyes perhaps an order of magnitude more expensive than several decades ago, old bottles are certainly worth testing.

The author tested old batches of dye, previously opened or unopened, against modern controls. He concluded that although some dye results are weak compared with modern controls, some yield the strength and spectra equivalent to contemporary samples. Dyes manufactured more than 100 years ago are certainly worth testing, as many such powdered compounds are extraordinarily stable. Titford gives a brief history, with references, of the industry and the various attempts to standardize colors.

Worth Experimentation

Coating with Tannic Acid and Iron

Ejima et al. (2013) have suggested a protocol that looks promising, especially when facing the following problem. For some cleared tissues there is insufficient contrast to see much detail when using a mounting medium with a high refractive index. On occasion, conventional dyes may not differentiate well, especially if specimens were exposed to chloral hydrate.

These authors devised an organo-metallic thin film that enhances contrast when using either LM or EM specimens. They combined tannic acid with ferric ions, which cross link and then bond to inorganic or organic surfaces. The solutions represent an organic ligand and an inorganic cross linker, respectively. They are safe and easily obtained chemicals.

Various concentrations of reactants were experimented with. A solution of each reactant is prepared. Upon stirring together, the two solutions appear to mix immediately and form a blue solution in a mildly acidic environment.

Concentrations. Tannic acid 0.40 mg/ml; $FeCl_3 \cdot 6H_2O$ range of 0.06 to 0.20 mg/ml. When introduced to specimens, the mixture forms a thin film, coating all materials tested including plastic, mineral or biological. Upon further observation it was found that the solution color is pH dependent: below pH 2, colorless; at pH 3–6, blue; in the alkaline range, red. At low pH (2 or below) the films disassemble, reversing the process when required. Refer to the original publication for explication of the physical chemistry.

CHAPTER 7

Mountants

This account describes mostly aqueous-based mountants, either temporary or potentially permanent. For an exhaustive listing and additional discussion, readers should see Brown (1997).

Aqueous-Based Mountants

Calcium Chloride

This solution at 20–40% strength is long-lived and has a refractive index (RI) similar to glass. The solution never evaporates and never hardens; and so, it must be stored flat. Preparations can be ringed with nail polish (see chapter 1 for discussion).

Five Mountants Suggested by J. A. Kiernan (1997)

Glycerol jelly. Gelatin powder 10 g; water 60 ml. Dissolve by warming. Add glycerol, 70 ml. For a preservative, add either of the following: (a) One drop of saturated aqueous phenol; or (b) 15 mg of sodium merthiolate. Warm to ca. 40°C before using. Also warm to remove bubbles. Discard when it becomes turbid.

Fructose syrup. Fructose (levulose) 15 g; water 5 ml. Heat to 60°C in oven 1–2 days until clear syrup forms. Supply is stable for several months. It does crystalize around the edges of the mount with time if cover is not sealed. An advantage of this is that the mount is easily taken apart if the section is to be handled again. The mixture is too acidic (pH 4.5) for preserving basic dyes. The author suggested adding a neutral buffer, but had not tried it.

Kiernan (1999), Van Brunt (personal communication; see chapter 4), and others have also commented on this mountant. The syrup, also sold as corn syrup or Karo, will eventually form a near-solid with time. It is unpredictable and sometimes too acidic for stain retention in specimens. If the syrup crystallizes, the slide can be immersed in water and the specimen recovered and remounted. As with other wet mounts, keep the cover and slide clean in enclosed storage. The cover cannot be wiped later if the mountant remains soft.

Apathy's gum syrup. Gum arabic (=gum acacia) 50 g; sucrose 50 g; water 50 ml; thymol (as a preservative) 1 small crystal. Final volume is ca. 100 ml. This is not easy to mix. Heat and stir the mixture in a water bath, but lumps of gum take days to dissolve. The final mixture keeps months at room temperature but should be discarded if sugar crystallizes or becomes moldy. Other preservatives, as listed in the section on glycerol jelly, may also be used. The RI of Apathy's is ca. 1.5, which is higher than most, therefore providing a transparency similar to that gained with resinous mounting media.

Polyvinylpyrrolidone (PVP). This mountant is a favorite of Kiernan (1990). Mix PVP (mw 10,000) 25 g with water 25 ml. Dissolve on a magnetic stirrer for several hours. Add glycerol, 1.0 ml, and thymol 1 small crystal. Polyvinyl alcohol can substitute for PVP.

Bottles of PVP mixture will last for several years at room temperature. The RI is 1.46, increasing as water evaporates at the edge of the mount. Therefore, for greatest transparency, wait several days before sealing the mount. The author recommends various additions if immunofluorescence is the goal.

This mountant is more runny than glycerol jelly or Apathy's, but it is easy to use and is usually free of bubbles.

Glycerin jelly. Kisser (1935) noted that this recipe has long been used for mounting a variety of materials. The suggested recipe and technique here tends to minimize the trapping of bubbles or other artifacts within the mount: gelatin 50 g; water 175 ml; glycerin 150 ml; phenol 2–3 g. Transfer the material from water to glycerin, and then to the slide, on which a small piece of glycerin jelly has been melted. Stir briefly and cover.

Hoyer's Medium

(Anderson 1954). Hoyer's medium is also discussed in chapters 2 and 5; it is included here because it clears coloration from bryophyte tissues. Chloral hydrate is the active clearing ingredient; it also has some clearing effect on vascular plant tissues. If there are dark deposits on your material, those must be initially cleared as much as possible or chloral hydrate will have an inadequate effect.

Mix ingredients, in order, at room temperature: distilled water 50 cc; gum arabic (USP flake) 30 g; chloral hydrate 200 g; glycerin 20 cc. Gum arabic goes into solution slowly. Flakes are best; crystals will work. Avoid powder, as it is difficult to work with. Use any available mechanical means of mixing. The mixture may seem to need filtering, but this is not necessary. After a few hours, bubbles will disappear and the solution will clear.

NOTES:

1. Before Anderson introduced the recipe to bryologists, Hoyer's medium was known to be useful for mounting mites, small insects, and fungi. The finished solution is very light yellow, and it has an indefinite shelf life in well-capped containers. The solution clears small plant parts and serves as a good permanent mountant for small, unstained whole mounts. Chloral hydrate destroys the effects of most stains, so this mountant should be avoided where staining is required.
2. Well-hydrated specimens of bryophytes or other thin plant materials should be transferred directly to drops of Hoyer's on a slide. Place cover glass carefully from one end to avoid trapping bubbles. Place flat on tray or work area.
3. Check specimen over a couple of days. At the cover-glass edge, a slight excess of mountant will evaporate slowly, forming a hardened margin. Away from the edges, the mountant remains viscous, so preparations should be stored flat.

4. Poorer grades of gum arabic may make a darker solution and perhaps leave sediment. Use the purest grade available.

Valap

(Tanaka, 1940). The name valap is an acronym derived from Vaseline, lanolin, and paraffin. It can be used as a cover-glass sealant. It has also been used to make chambers for observing live water-borne cultures. Hanging drops on a cover glass are placed onto a ring of valap and pressed lightly to seal. In use, a flask containing this sealant is heated gently on a hot plate. When melted, a thin glass rod is used to apply the material to a slide, or edge of a cover glass. The stock flask can be reheated as often as necessary.

To make valap. On a hot plate, melt together: Vaseline 2 parts; lanolin 2 parts; paraffin 1 part. Store the mixture in an Erlenmeyer flask with a cotton plug or loose cap. When cold, valap is plastic with a consistency of softened paraffin.

Making Aqueous Slide-Mounts Permanent

(McCrone, 1999). An aqueous mount should be dry at the edges of the cover glass. Blot or wick away excess water at the edges, and wait until any moisture on the slide surface and cover-glass edges evaporates. Various sealants can be fed around the edges of the cover. The following materials work; all should be as fluid as possible:

1. Nail polish, colorless or not, thinned with amyl acetate if necessary.
2. Aroclor 5442: add to slide when both slide and Aroclor are at 60°C.
3. Canada Balsam, room temperature, diluted with organic solvent such as amyl or ethyl acetate, carbon tetrachloride, etc.
4. Epoxy cement, slow-curing type.
5. Vaseline, especially for surrounding living cells. With a spatula prepare a shallow, square enclosure of Vaseline ridges. Add an aqueous suspension of culture or pond water with bits of plant material. Mount the cover, lightly pressing it into place over the Vaseline walls.

Some air should be trapped within. Living activity can be observed over weeks or perhaps months. The enclosure can be sized for large, 22×40 mm covers, or smaller ones, when preparing a long-lasting protozoan or algal microaquarium.

NOTE: Valap is even better for this purpose.
Each preparation above should be inspected after 20–30 minutes to be sure it remains sealed, especially when using nail polish.

Making Resin-Mounts

Silverman (1999), among others, has noted that there is no longer any need to use the xylene or toluene ring solvents for clearing or mounting. D-Limonene is a citrus oil, an aliphatic hydrocarbon (monoterpene) distilled from the peel as a byproduct of juice production. It has a low flash point and lower volatility, advantages when coverslipping.

Limonene is rated as generally regarded as safe (GRAS), although some workers, rarely, have reported becoming sensitized to its odor. Brands such as Citra Solv and Histo-Clear, among others, are available. Histomount is an example of a resin-mount using the limonene solvent.

CHAPTER 8

Microscopy and Measurement: The Light Microscope

It is essential that structural botanists use the optical or light microscope (LM) carefully and competently. This sophisticated instrument evolved over 200 years, but it takes only a moment to render one nonfunctional. There aren't many courses on the effective operation of a light microscope today.

While light microscopes of different brands and vintages may look quite different, clinical or research-grade models have a basic array of lenses and controls that you should be able to identify. Master their operations, which include (a) Keeping the body (stand) clean and functioning; (b) Keeping the optics clean and functioning; and (c) Keeping the illumination system clean and aligned.

Light microscopy can be a complex subject, and this primer is not an attempt to substitute for serious manuals that unveil its subtleties. Major manufacturers have published introductory guides to their products. Other interesting sources covering microscope operation are Burrells (1977), Smith (1990), Slater and Slater (1992), and Rost and Oldfield (2000). Some microtechnique guides mentioned in the introduction of this book also cover microscope use.

Among light's advantages is the fact that different wavelengths of the visible spectrum are perceived as different colors. Therefore, specimens are regarded as *amplitude objects*. By the use of dyes and filters, one can emphasize different structures within the same tissue. LM's principal limitation is that resolution is limited by wavelength: the diffraction barrier. This is a moot point in morphology and anatomy, where lower magnifications are usually required.

Parts of the Light Microscope

From the glossary below, identify and know the following terms.

Stand. The large metal casting that supports all of the other parts.

Head. On research-grade microscopes, the head is a housing near the top of the stand that holds at least a pair of ocular lenses. The pair is adjustable for interpupillary distance, and one ocular mount allows adjustment for variations in human eye focal distances.

The best heads are trinocular. They have a third tube and an ocular for the purpose of attaching a camera. While not essential, it is hugely convenient to have a trinocular head if photomicrographs are to be made routinely.

On student stands or older research microscopes, the head is a cylindrical tube that encases the optical axis of the microscope. One ocular lens is on the upper end of the tube and the objectives are on the lower end. Modern ergonomic heads have folded optical paths. Prisms bend the vertical optical path to one angled toward the eyes.

Oculars. Also called *eyepieces*. On research or clinical-grade microscopes, oculars occur in pairs. They are usually removable lenses that human eyes look through to see the microscopic image. They are available in different magnifications such as 8X, 10X, 15X or 20X. However, for enhanced resolution of detail, use a higher *power* objective lens, not a higher magnification ocular lens.

Objectives. There are usually several such lenses on a rotating nosepiece at the bottom of the head or tube. They interface with the glass slide bearing the prepared specimen. The design of the objective lens determines most of the light-gathering power, resolution of detail, enlargement, and clarity of the specimen image. They are available in steps from 1X to 100X. Total image magnification is determined by multiplying the power of the ocular times the power of the objective lens. These figures are always engraved on lens bodies.

Also engraved on the lens body is another number: the numerical aperture (NA). This is the angle of the cone of light that the lens is capable of capturing. The higher the number, the greater the resolution of detail.

Stage. The flat platform that holds the microscope slide. It separates the objective lenses from the condenser below. A slide holder is usually attached to *stage drives*, allowing movement of the specimen slide on x and y axes.

Substage. An adjustable lens and diaphragm system that collects and focuses the illumination beam through the specimen, where it then enters the objective lens. Usually two adjusting screws are present that allow the light path to be centered.

Focusing knobs. Either the nosepiece with objective lenses, or the stage, is moveable, up or down, relative to the other. The larger knob is for coarse adjustment; the coaxially mounted smaller knob is for fine adjustment. The substage is also focused by means of a separate knob beneath the stage.

Substage diaphragm. An iris diaphragm that controls the angle of the cone of light that enters the objective lenses. The more open the diaphragm, the greater the resolution of detail, the brighter the image, and the shallower the depth of field.

Field diaphragm. An iris diaphragm found between the lamp housing and the substage. It controls the diameter of the illumination. It serves as a point on which to focus the light source in Köhler illumination (described below). By the use of two substage adjusting screws, this diaphragm can be moved to center the light source (optical path).

Illumination. In the rear of the stand, or integrated into its base, is a lamp housing. Traditionally, its lamp is a tungsten or tungsten-halogen bulb, although other types can also be used. The bulb is the distal end of the optical path that extends, on the proximal end, to the retina of the human eye. Brightness is usually controlled by a transformer and a voltage rheostat. Follow the manufacturer's recommendation for optimal setting for longest bulb life.

At the time of this writing, light-emitting diode (LED) lamps are also becoming available. See discussion below.

Lens Care

As with any lens system, the microscope lenses should be kept clean and dust-free. Cover the instrument when not in use. A lens brush or lens cloth can be used sparingly on the outer lens surfaces. Be sure that the removable ocular lens is always in position to prevent dust from accumulating on the inside surfaces of the prisms or objective lenses.

The upper surface of the ocular lens, which picks up dust and oil from eyelashes can be cleaned with solvents such as Windex glass cleaner (S. C. Johnson & Son, Racine, WI) or the lens cleaners supplied by opticians.

Cleaning the Bottom of the Objective Lens

Usually, it is the exposed surface of the objective lens that is most vulnerable to picking up more resistant oils and dust. Visual clarity absolutely requires that this surface be clean. Therefore you may need to use a chemical solvent applied with a lens paper. It should be remembered that exposed surfaces of the objective lens are more easily scratched when compared to some forms of hardened glass. Anti-reflective coatings, when present, are also vulnerable to damage.

Duke (2004) recommends beginning with an air-duster to blow off dust, the most common problem; but do not use a brand that may leak aerosol propellants. Do not use lens brushes, which may accumulate oils over time. General-purpose papers such as Kim Wipes (Kimberly-Clark Corporation, Irving, TX) are suitable for stages and mechanical parts of microscopes but not for lenses. Lens papers should be used. However, hard-surfaced ones should not be used to rub the lens elements. The following method is recommended as it cleans the lens surface using no pressure, thereby avoiding damaging the glass surface:

1. Remove (unscrew) the objective lens from the microscope and place it upside down on a clean table surface. Even better, place the upside-down lens on the clean stage of a stereo microscope so that you can examine your cleaning progress. Or, have a hand lens available to note your progress.
2. Place a piece of lens paper on the lens surface, add a drop of cleaning solvent, and drag the paper across the surface. This causes the solvent to roll around as the paper drags across. This procedure is pressure-free and reduces the amount of time the solvent is in contact with the lens. Excess solvent application must be avoided.
3. To remove a resistant surface contaminant, such as a bit of cover glass cement, fold a sheet of lens paper into a pointed probe, dip it into the appropriate solvent, then carefully work the area of the deposit. Avoid flooding the objective lens with solvent.

CAUTIONS: For reliable results—that is, to avoid disaster—it is imperative that you contact the microscope manufacturer for recommended cleaning solvents. Among the recommended solvents, especially for cleaning oil immersion lenses, are acetone, ethyl ether, carbon tetrachloride, methylene chloride, naphtha, and xylene. These are definitely not interchangeable. Leitz, Zeiss, and Japanese manufacturers, for example, each have very different recommendations based on the vulnerability of the cements the companies use to seal multi-element lens systems. The wrong cleaning solvent may get beneath the outer lens. This may partially de-cement the lens element stack, thereby rendering the objective useless.

Chiovetti (2001) noted that some recommendations for lens cleaning can severely damage multi-element objective lenses. Especially mentioned as inappropriate is sonicating in acetone. This could be the fastest way to dissolve the sealants and lens element cements. The key is to use quickly evaporating solvents, sparingly enough that they evaporate before penetrating sealants.

Illumination Techniques

First, a Couple of Essential Principles.

Diffraction. Light consists of waves of various lengths that we see as different colors. Because of its wave nature, light can bend around objects or sharp edges. When light waves pass a sharp edge, they generate a new wavefront at that edge. The new wave has been retarded, compared to adjacent waves that missed the object entirely. When the crests of these two adjacent waves combine, the result is *constructive interference*. When the crest of one wave combines with the trough of another, the two waves cancel each other out, which is *destructive interference.*

When the condenser is lowered well below the specimen, thereby lowering the numerical aperture, the visual effect is a pattern of light and dark circles around objects in a microscopic field. These are the constructive and destructive wave fronts. The higher the numerical aperture of the condenser (below the specimen) and the objective lens (above the specimen), the less obvious the diffraction phenomena and the sharper the detail. The greater the development of diffraction patterns, the greater the loss of detail in the field of view.

Refraction. Frequently, light will pass successively through two different transparent media. Air, for instance, is a less dense medium than glass. At that boundary, light passing through air into glass, at an angle, will bend in a direction more perpendicular to the surface of the glass. Köhler Illumination, discussed below, will minimize the development of optical artifacts due to diffraction or refraction that could adversely affect your interpretation of your specimen preparations.

Köhler Illumination

Your goal in basic LM is a clear specimen image, with appropriate contrast, against a bright, evenly lit background. Published photomicrographs that don't meet this goal are still too common and reflect poorly on the researcher's competence. The default position of the illumination path is called Köhler illumination. Your knowledge of the microscope's proper alignment is essential before considering several useful off-axis techniques to be discussed below.

Köhler illumination is a form of *critical illumination* that is considered to be the best method for aligning the microscope and the specimen's field of illumination. Its purpose is to provide the highest resolution of detail of which the lenses are capable, as well as a uniformly illuminated background. When the conditions of Köhler illumination are met, the rays of light are parallel as they pass through the specimen. There are other ways to use the microscope, but Köhler illumination should be your natural starting point.

Alignment

The following steps pertain to microscopes having lamps with filaments that can be focused and properly centered. See below for discussion of LED illuminators.

1. Turn on the lamp. Open wide the field and substage diaphragms.
2. Rotate the 10X objective lens into place. Place a specimen slide on the stage and focus on it.
3. Close down the field diaphragm. Turn the knob on the substage assembly up and down until it focuses on the small opening of the field diaphragm. Both the specimen and the field diaphragm should be in focus. A small circle of light is superimposed on the specimen.
4. Open the field diaphragm until its edges approximately conform to the edge of the visual field but remain slightly visible. Looking at the stage from the side, the top lens of the substage condenser should be just below the level of the specimen slide. The specimen and field diaphragm edges should both be in sharp focus.
5. If the field diaphragm circle is off-axis, or not centered with the visual field, use the two adjusting knobs on the substage to bring the two into alignment. These knobs are at 90 degrees to each other, forming an *x-y* axis. Now open the field diaphragm so that it is just barely visible at the edge of the field.
6. While looking at the specimen, close the substage diaphragm slowly until the specimen begins to darken. If you take out one of the oculars and look directly down the tube, you should see the edges of the substage diaphragm. Under normal conditions, you want to avoid having

the substage diaphragm closed further than where you begin to see this dark ring at the edge of the view. Some closure is often helpful to provide more contrast, especially with light-colored specimens. Too much closure of the substage diaphragm increases diffraction and degrades the image. For maximum resolving power, the objective lens must include the widest cone of light for which it was designed.

This completes the basic set-up for Köhler illumination. With each change in objective lens power, the diaphragm diameters must be reset, but the illumination alignment should remain unchanged. These settings put the diameter of the illuminated field under the control of the field diaphragm and put contrast and resolution under the control of the substage diaphragm. With practice, the set-up can be checked and touched up quickly at the beginning of a session. For photomicrography (see below), good illumination is especially important.

Brightness and Filtration

The visual image should be pleasing to the eye and not so bright as to cause eyestrain. There are two methods to control brightness. Do not use the substage diaphragm to do this.

Method 1. Use the rheostat knob to lower the voltage on the transformer, but follow the manufacturer's instructions for the recommended settings. A problem with changing voltage settings is that the color temperature changes, and this affects the color balance of your photographic image, which is mostly a problem if using color film. With digital images this can be adjusted later using in-camera settings or computer applications.

Method 2. The preferred way, especially for photomicrography, is to use neutral density filters. These filters reduce the light intensity uniformly across the visible spectrum. Some microscopes have such filters built into the optical path, or they can be added to a slot in the substage.

NOTE: Neutral density filters can be easily made by using developed, unexposed black and white film. Such film is light gray, which is called the "film base plus fog." You can use the leader of a roll of images. You can mount negative-sized pieces in a slide mount and place one or more on the microscope substage. Filters of various types can be used for enhancing contrast or for color correction. The color of your specimens, called *amplitude objects* in microscope theory, can be strongly altered. A green filter, for example, will lighten green areas of the specimen and darken or blacken red areas. Your goal is good contrast without loss of detail. Experiment here with various available filters. The color temperature of the lamp, plus the sensitivity of your film or digital sensor, can vary from one microscope to another (Joubert and Sharma, 2011). A side effect of filtration is a more evenly illuminated background in images made with digital sensors.

Polarizing filters are quite useful for detecting birefringent crystals but are of limited application for general plant morphology. The polarizer filter fits beneath the condenser and the analyzer above the objective lens, often in a slot or other dedicated position. To identify crystals, one needs a petrographic microscope with a rotatable, scaled, circular stage, or other means for measuring the angle of rotation of polarized light from the extinction position. Birefringent crystals, such as the different hydration states of calcium oxalate, are easily detected by their bright presence against a dark background. See texts on microscopy for essential details on this topic.

Diffracted Light Contrast (DLC)

Usually, improvements in resolution are only found in expensive optical systems such as phase contrast, differential interference contrast (Nomarski DIC), Hoffman modulation contrast, or single side-band contrast. These techniques involve sophisticated modifications of the illuminating beam. Piekos (1999, 2003, 2006) has described a simple way to get the best advantages of off-axis illumination. These include enhanced contrast, as well as resolution previously thought to be beyond the capability of student or clinical microscopes. This technique works especially well with thin, unstained specimens such as epidermal layers.

Initially the microscope is set up for Köhler illumination. The technique involves insertion of an edge plate on a flat surface between the field diaphragm and the condenser. This puts it in focus with the field diaphragm. The edge plate is made from thin metal, anodized or painted black. It is ca. 1.5 cm long by 1 cm wide, with three straight sides. The fourth, long, side is convex, its curve having a radius of 32 mm. On one of the narrow sides a small 2- to 3-cm-long handle is attached, such that the whole device resembles a paddle with one convex side. This can also be achieved by placing a cut piece of black tape on a microscope slide.

The shape of the edge of the opaque barrier is important. Convex edges give higher contrast than straight or concave edges. Piekos believes this indicates that diffracted light, rather than oblique illumination, yields the enhanced contrast. Because the plate is below the condenser, the condenser's full numerical aperture can still be used. This technique can even be used with very simple condensers that lack adjustable apertures.

IN USE: The edge plate is placed in the optical path so that the convex edge is in focus with the specimen. Then defocus (lower) the condenser until a broad band of diffracted light is superimposed on the specimen. Unstained specimens that lie in the light-dark gradient created by the edge of the plate show higher contrast and resolution.

Oblique Illumination

Abramowitz (1987, p. 17) suggests placing a stop in a filter holder below the lower lens of the condenser. This can be a clear glass disk where half of the light is blocked out by tape. Light passes through the specimen from one side of the condenser, thereby displacing the zeroth order to one side of the objective.

Theory. Diffracted orders (at least first order) are also included from the specimen to the other side of the objective. Diffracted orders on the other side of the specimen miss being included in the objective because of the zeroth order obliquity. The specimen is therefore viewed in pseudo-relief, giving it a three-dimensional clarity.

Circular Oblique Lighting (COL)

This little-known technique has been around for many decades. Among other names, it has also been referred to as "hollow cone illumination," "annular bright field illumination," or "conical illumination." Considered to be better than phase optics, the method has the merit of being a virtually cost-free way of improving contrast and resolution for thin subjects, such as pond life and epidermal preparations or peels. Clarke (2012) and James (2012) have detailed online articles on the history as well as various approaches to modifying microscopes.

The technique resembles dark field microscopy in that the zeroth order beam is blocked from contributing to the image. In this case, the specimen background is illuminated to a uniform gray. A circular stop, similar to a dark field stop, is inserted into the optical path close to and above the field diaphragm. The COL stop is smaller. James suggested placing a thin, 10–15 mm diameter disk on the substage glass.

My own experiments using a Leitz Ortholux microscope (Leica Camera, Wetzlar, Germany) indicate that $^{1}/_{4}$-inch or $^{5}/_{16}$-inch disks work best. To work well, the disk should be accurately cut from very thin opaque material. It is easiest to use a hand-held paper punch to make circles in aluminum foil of these diameters. Support the foil with a piece of paper or index card to make a clean cut.

Procedure

1. Place a specimen on the microscope stage and focus. See that the optical path is centered and Köhler illumination is achieved. Open the field and substage diaphragms.
2. Place the opaque disk on the substage glass and center it.
3. Lower the substage condenser until a dark circle is superimposed over the specimen. This confirms the disk's centrality or the need to center it.
4. Rack up the condenser until the dark zone disappears. Sharp-edged cell walls should show up against an illuminated, uniform, but contrasting background.
5. The brightness of the field can be controlled by closing down the field diaphragm.

This technique is especially effective when using an Abbe condenser because it does not focus all light rays to the same point. But it also works well with any type of condenser. Apparently, the pores on diatom frustules can be resolved at numerical apertures below those found in oil immersion lenses.

Light-Emitting Diode (LED) Illumination

This fast-developing technology obeys Moore's Law and will probably replace conventional microscope lamps. Among the advantages of LEDs are low cost, very low energy draw, low operating temperature, and very long life. They are semiconductors based on the element gallium. The movement of electrons produces photons through the process of electroluminescence. By combining gallium with various salts and other elements, LEDs are now manufactured to be capable of emitting portions of the entire visible spectrum. This makes them ideal replacements for the energy-hungry and heat-generating "mercury burners" traditionally used in fluorescence microscopy (Beacher, 2011; Foster, 2011).

For conventional LM, white light LEDs are manufactured in a ring-light configuration, or as a light plate that can be placed anywhere below the substage.

When using LED illuminators, Köhler adjustments are much less important because there is no single filament that must be centered and focused out of the plane of the specimen. An unevenly illuminated background or specimen-shadowing should not be a problem. Reasonably priced examples can be seen at www.amscope.com. Their two types produce white light and are available in square or circular configuration.

Measuring Leaf Areas

For those who lack access to expensive leaf area measuring machines, the following tools will do the job.

Polar Planimeter

Kappelle and Leal (1996) measured leaf area using a polar planimeter. After tracing the outline of a leaf on paper, the polar planimeter then traces the drawn outline to compute the leaf area. This is a fast technique. Using the operating instructions that come with a particular model, the measurement wheel must first be calibrated. This tool can also be used on topographic maps to measure, for example, areas occupied by vegetation types, or other features, with high accuracy.

These authors also used the Raunkiaer Leaf Class System. Leaves are classed as nanophylls (0.25–2.25 cm^2), microphylls (2.25–20.25 cm^2), notophylls (20.25–45.0 cm^2), mesophylls (45.0–82.25 cm^2), and macrophylls (182.25–1640.25 cm^2). This system is still favored by many botanists. See Webb (1959) for further description.

I also use a polar planimeter to measure leaf area. If purchasing a planimeter, choose a model with an adjustable tracer arm such as the Lasico model series 30 (Los Angeles Scientific Instrument Co., Inc., Los Angeles, CA). The adjustable arm makes it possible to get a direct readout without having to calculate a multiplier constant.

Calibrate the planimeter by drawing a box of a known cm^2, or a circle of known cm^2. These should be of the size class for the leaves you are measuring. Adjust the tracer arm length to the vicinity of 18 cm. Practice tracing the test areas until accurate repeatable results are obtained.

By trial, I found the best repeatable results as follows: Obtain a piece of tracing (translucent) Mylar and lay it with the matte (rough) surface up. With blue masking tape, fasten the corners of the Mylar to a smooth flat table top. Size it generously so that the leaf or leaf drawing is inserted beneath the right edge of the Mylar. The planimeter always rides on the left side within the margins of the Mylar. The drag of the measuring wheel on the Mylar's matte surface seems to ensure an accurate reading.

Dot Planimeter

Dolph (1977) reviewed a number of techniques for estimating leaf area based on measurements of the length and width of a leaf blade. Such schemes involve multiplying length by width, and then by a correction factor such as 0.667 intended to account for base and apex taper. However, he noted that these approaches fail to be accurate when the investigator is faced with a sample that shows large variations in leaf shape. Leaf lobing, for instance, is common in temperate woody floras.

Dolph recommended making a clear plastic template marked with spots or small holes, spaced 5 mm apart, on a grid of vertical and horizontal rows. The template, called a *dot planimeter*, can be laid over a leaf or leaf tracing. All holes found within the leaf borders are counted. On one half of the leaf, count any dot that touches the margin; on the other half, do not count the holes just touching the margin. Each included hole is the center of 25 mm^2. When this figure is multiplied by the total number of holes over the leaf, you have an accurate measure of leaf area. I found greatest repeatability by placing the planimeter over the leaf outline so that the rows of dots are at a 45-degree angle to the direction of midrib. In design, one could use tighter or wider spacing of holes depending on the size range of the leaves being measured.

NOTE: Depending on how the dot planimeter is placed on a leaf, there may be as much as several tenths of a centimeter difference in repeatability. The same is true of the polar planimeter. With practice, accuracy increases; I found the estimations to be within reasonable limits for ecomorphological studies.

Counting Mesophyll Structures

Sundberg and Zahn (1985) made thin hand-sections of leaves and examined them using light microscopy at a magnification of 675X A 20×20 grid reticule was placed in the ocular lens for point-counting analysis. The authors made separate counts for the number of points occurring over chloroplasts, vacuoles of cells containing chloroplasts, intercellular spaces, and non-photosynthetic cells. To assure that an optically thin section was sampled, only chloroplasts in sharp focus were counted.

They derived an index of morphological mesophyll succulence ($=S_{mm}$). This is defined as the volume density of vacuoles/volume density of chloroplasts in the photosynthetic tissue. It is calculated as the ratio of point counts for these structures.

Counting Stomata per Area

(Keating). An excellent tool for counting objects per area in a light microscopic field is the Whipple disc. This is an ocular micrometer that is placed in one of the microscope's oculars. This device, little known among plant anatomists, has been used traditionally for counting mineral particles in coal or rock sections, as well as for counting aquatic microorganisms such as protozoans or algal colonies.

My disc is etched with a square grid, 8 mm on a side, ruled across and down, 10×10, to produce 100 squares. One square, just off center, is further subdivided into smaller boxes that provide a reference point. Visually, the whole grid is easily subdivided into quartiles, each with 5×5=25 squares.

At the outset, calibrate the dimensions of the squares with a stage micrometer for each objective lens you intend to use, then derive the area of a quartile (in micrometers). Divide this into 1 million, which is the number of square micrometers in a square millimeter. This yields the multiplication factor for your sample area.

An Example

Leaf portions are cleared and mounted on slides. The epidermis to be sampled is focused and the condenser lowered, as necessary, to provide contrast and better visibility. I anchor a stomate in a reference box, which is the corner of the sample field, or quartile. Next, record all of the stomata in the sample field. At the left field edge, stomata more than

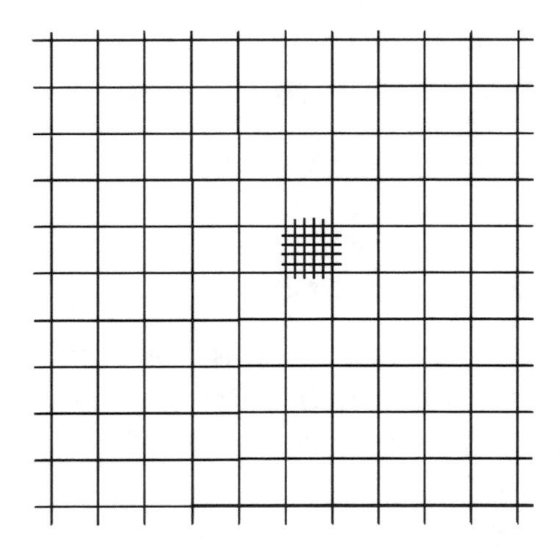

FIGURE 6. The Whipple disk is placed within an ocular lens at the plane of focus. To calibrate it, focus on a stage micrometer using the required objective lenses.

half in the field are included in the count; at the right field edge, they are not. With the stage drive, move the reference box to a new random sample field and repeat the process.

I count 10 fields per leaf and derive the average number for the 10 samples. The 25X objective, best for my material, usually yields between 3 and 10 stomata per sample field. In my case, for example, when using the 25X objective, I multiply the average stomate count per sample field by 32.6, which yields the number of stomata per square millimeter. I have tried larger and smaller sample areas and found the regimen as described to be highly repeatable. With practice, sampling a leaf accurately takes about as much time as required to read this account.

NOTES:

1. Microscope objective lenses of 16X and 25X are really convenient for micromorphology observations. The 25X lens, yielding a total magnification of 250X, works well for many species when counting finite numbers of stomata in a grid field. Powers of 16X or 40X may work better for other taxa.
2. For the clearest view, do one of three things, being first sure the field aperture coincides with the field of view and is in focus: (a) Lower the condenser to provide more depth of field, or contrast; (b) Close down the substage aperture; or (c) Try using the oblique or circular oblique techniques noted above. The first and third options give the most uniform illumination with my Leitz Ortholux, but this is not always the case with other makes.

Alternative Stomate-Counting Methods

It may be that a grid ocular micrometer is unavailable, or counts involve totaling all cells in a field to derive a stomatal index (SI), or counting stomata in a grid framework may cause eyestrain. In these cases, try any of the following methods.

Method 1. Make an image of the field (film or digital), and of a stage micrometer at that magnification. Calibrate the diameter or radius of the field and derive the area of a circle (πr^2). Cells or apertures are then counted from a print or transparency.

Method 2. Negatives or transparencies can be projected using an enlarger or slide projector. The same can be done with a digital projector. An image of the stage micrometer is first projected on the screen to calibrate the image size and area to be counted.

Method 3. The image of the stage micrometer can be observed on the computer screen, followed by images of the specimens. Cells can be counted at the screen.

Method 4. Use image analysis technology, such as ImageJ, and other programs. If the digital image is on the monitor screen next to the microscope, one defines the object to be counted. This only works if the stomata stand out well from adjacent cellular background structure. See also Kubinova (1994) for a discussion of stereological and random sampling methodology.

Optical Enhancement of Cellular Features in Wood

(Keating, unpublished). To assist a colleague, I was asked to identify a 5000-year-old conifer wood sample from a Michigan bog. Using a razor blade, I took the piece, still wet from a core sample, and cut thin slivers by hand to represent cross, radial, and tangential views. The sections were mounted in the high-RI calcium chloride solution and observed under the light microscope.

The unstained wood sections were the usual tan color and showed the expected amount of deterioration. Features such as cross-field pitting and tori were impossible to see. Embedding and microtome-sectioning might have been an option but for the presence of mineral soil particles that would quickly have destroyed the blade edge.

To enhance visibility, I placed an FP 490 (bandpass) interference filter in the holder beneath the condenser. The change was dramatic. The wood cells were now red against a neutral blue-gray background. Tori of tracheary pits were easily detected, as was cross-field pitting and all other diagnostic features.

NOTE: A strong light source, such as tungsten-halogen, is necessary for this procedure because bandpass filters reject most other light wave frequencies. With ordinary thin sections, the results are unimpressive; but with tan wood hand-sections it's a different story.

Mitigating Vibration Problems

(Moss, 1999). Some microscopists have reported it to be nearly impossible to get sharp photomicrographs because of building vibration. Those who use other kinds of sensitive equipment also report such problems. Even in basement laboratories, subways, truck or public transport traffic, or vibrating building mechanical equipment cannot be controlled. Try one of the following methods to mitigate the vibration.

Increase the exposure index. With digital cameras, try setting a high exposure index such as EI 800. This will not noticeably degrade the image, but will instead result in a high shutter speed. The result will "stop action" and provide a sharp exposure.

Use film with lower EI. Films with an EI (=ASA) at 400 or above may also solve the problem. If it is necessary to use film with a lower EI, there is still hope.

Use a vibration damping table. Vibration damping tables are available, or you can make your own. To make a microscope table of about 100–250 pounds, place a heavy piece of plywood beneath the table legs. Then, various investigators have tried the following: (a) Place inflated inner tubes from bicycle tires under the heavy board; (b) Place tennis balls at the corners of the board. To restrain them from rolling, make depressions by drilling shallow 2-inch diameter depressions with a Forstner bit, or hole cutter; (c) Obtain silicon elastomer shock absorbers, such as Sorbothane vibration isolation mounts (Sorbothane Inc., Kent, OH); (d) Extra mass is good. A heavy piece of granite or ceramic countertop may be placed under the microscope. Sand or lead shot has also been used to provide mass.

NOTE: There is no "one size fits all." The amplitude of sound or physical vibration resonances will be different in each situation, and so require experimentation.

Photomicrography

Digital Capture

Contemporary photomicrographic equipment is digital, frequently dedicated to a brand of microscope, and frequently linked to a laptop computer. The cameras are based on image capture with a charge coupled device (CCD) or with a complementary metal-oxide semiconductor (CMOS). The CCD detector is said to be slightly more sensitive, but it produces somewhat more noise. Top-grade cameras for LM photomicrography are based usually on the CMOS technology. Presently, these devices have the edge over a range of magnifications, signal-to-noise ratio over a range of exposure times, and overall image quality (Joubert & Sharma, 2011). They are also alleged to use about one-hundredth the power draw of a CCD.

Adapting Digital Equipment to Existing Microscopes

Companies such as Martin Microscope Company (Easley, SC) and Scopetronix make microscope adaptors for several digital camera models that they believe make the best match to the microscope. In the lower price range, digicams have non-removable zoom lenses, but some of these can be mated to a microscope. It helps if they have a screw thread so that adaptors or lens tubes can be attached. At the next level are camera bodies with lenses that can be removed. In either case, microscope dealers, makers of adaptors, digital camera vendors, and satisfied or dissatisfied users are good sources of information.

Megapixel density is important and should be above 5, but numbers above 8 are not required. A fairly high-density image (such as "fine") is better for later cropping and manipulation. Image quality should be high file size in either TIFF or JPEG format. The latter images are regarded as "lossy"—that is, pixel information is lost and images develop "noise" pixels if the files are repeatedly opened and closed. This turns out not to be a problem at modest compression ratios, such as 4:1. Therefore, try using JPEG images in combination with a larger file size.

Images made at the highest file size, such as RAW, may not be a good idea for two reasons. First, the huge file sizes devour storage space, and such images can be slow to download and manipulate. Second, check to see if your image processing program can open RAW files. Not all are capable. In my opinion, even TIFF files are larger than necessary, and I have gotten good publishable results using JPEG files.

NOTE: The above discussion assumes final print sizes as normally found in published papers. For making large exhibition prints, image file sizes need to be large. See www.kenrockwell.com for a discussion of file size and storage.

Background Illumination

The problem of ending up with an unevenly lit background is especially serious with digital capture, even when using Köhler illumination. Even though your eye, or film, registers an evenly lit background, the CCD or CMOS chips are hypersensitive to slight amplitude variations that can magnify and record background gradients. This detracts from the quality of your final image. George Yatskievich (personal communication) suggested the following technique for making high-quality black and white (monochrome) images:

1. Place a green filter in the substage filter holder. These are usually marked as having a 546 nm maximum transmittance. The bell-curve transmittance range is usually about 500–600 nm.
2. Photograph in color as usual.
3. Call up the file in an image manipulation program such as Photoshop or the free downloads Picasa or GIMP.
4. Change mode to gray scale, which deletes color information.
5. Adjust brightness/contrast to provide a brighter, clearer image. It is always necessary to increase contrast when converting a color image

to black and white, since the dynamic range of the original color capture is always narrower than the potential gray scale. (For those familiar with the "zone system," the usual black and white range is nine zones, as opposed to seven zones for color.) Also, some contrast may be lost during the publication process.

Part of the reason for unevenly illuminated backgrounds is slight polarization effects. These are present to some degree in most microscope lenses due to *optical strain*, which is caused by slightly uneven pressure on mounted lens elements. For quantitative polarization, microscopes with strain-free lenses are available at much higher costs. Green filtration evens out the background by attenuating other conflicting wavelengths and amplitudes.

Making the Digital Photomicrograph

None of the following applies for those with access to the new generation of integrated automatic light microscopes. These units have built-in scale lines and many other amenities. This is a rapidly emerging field where you must follow the explicit manufacturer's instructions.

My technique, as follows, assumes a trinocular microscope with an attached digicam. The image may be stored on a memory card, or the camera may be wired directly to a laptop computer. Straight photomicrography without special techniques is outlined below.

Procedure

Microscope.

1. Remove all polarizers.
2. Use ND filters for light control.
3. Use diffusion filter, or green filter for background control.
4. Adjust substage diaphragm at the binocular head and leave alone while composing and focusing the image at the camera.

Camera.

1. Remove 10X eyepiece from third ocular tube if the camera body has its own lens.
2. Use a house current power adapter for the camera, if it has one, rather than batteries.
3. Focus: Change camera from automatic to manual. Focus using the microscope's fine focus control rather than the camera's automatic focus.
4. Flash: Cancel.
5. Pixel density: Normal or fine, but usually not RAW.
6. Scale information: Photograph a substage micrometer for all objective lens/ocular combinations that you will be using for images. Cross-index this with your computer file numbers later.
7. Zoom: Set at maximum, in order to fill the frame with the image.
8. Focus aid: If necessary, use the camera's electronic telezoom (electronic magnification) to focus on part of image using the fine focus knob of microscope. After focusing, return to the highest optical zoom setting.
9. Make exposure.
10. Record. Always use a photo log sheet or notebook to note specimen and conditions used.

Image Manipulation

Transfer file, unaltered, from flash memory to computer. This process usually goes faster when using a memory card reader rather than downloading from the camera itself.

At the computer: (a) Purge bad shots (if not already done at the camera); (b) Bring up the new file; (c) Using your image data log, record or change file names or image numbers and reconcile these accurately against your exposure log; and (d) Rotate all images to right side up.

On the computer, rotate or discard images as necessary. If archival storage is planned, there are several options. Some feel that an external hard drive is the most stable option, and 500 gigabyte drives are getting less pricey. High capacity flash drives such as those from PNY Technologies (Parsippany-Troy Hills, NJ)—above 8GB storage—also work for smaller collections of images. I avoid brands, such as SanDisk Corporation (Milpitas, CA), that try to load extraneous software onto my computer. Alternatively, one can burn a CD or have it done at the drugstore or camera shop. If done commercially, one can obtain a set of index photos to go with the CD.

Using Adobe Photoshop to Make Grayscale from Color Photomicrographs

1. Adobe → Open → Click on file number to load file into Photoshop.
2. At task bar: Image → Size → Width 5″ → Resolution 300
3. Mode → Grayscale
4. Image → Brightness/contrast: adjust
5. File → Save as: [image name or number].jpg, medium file
6. Save to archive file, flash drive, or new CD.

Film Capture

For most publication or presentation purposes, a 35-mm camera, if you already have one, remains the most cost-effective option for documenting microscopic images. It provides superior resolution in color or black and white. I measured film grain detail using Kodak Tri-X (Eastman Kodak Company) and determined that this film provides the equivalent of about 13 megapixels of digital capture. (There is some variation based on film speed, and see below). As this is written, the buyer's market has never been more favorable and some excellent film equipment can be obtained for shockingly low prices. Internet auctions or larger camera stores have great bargains.

For electronic presentation or preparing electronic files for publication, slides or negatives can be inexpensively scanned as required. When you have your slides or negatives processed, you can order a CD with your digital files for a modest price.

One major advantage of film negatives is their archival permanence. My own archive demonstrates that well-processed negatives older than 100 years have lost no information content. The jury is still out on the reliability of contemporary digital storage media. The most recent estimates I have seen are not encouraging.

Mating cameras to microscopes. While dedicated photomicrographic camera bodies are ideal, one can adapt others as follows. Typically, an ordinary single lens reflex (SLR) camera body, without a lens, is mounted on the third ocular tube of the microscope. The microscope's ocular lens forms the final image for the camera. An extension tube is mounted on the camera body, which is in turn mated to the microscope tube. For best results, only the camera's body is placed over the photomicrographic ocular. Placing a camera lens over the microscope ocular results in too many lens elements, and therefore a degraded image.

For help with equipment, seek out the used parts bin of an older photography store in a large city. Very useful equipment can be obtained, especially with the help of an older salesperson who grew up with this technology. For years, my students and I used a Nikon body (Nikon Corporation, Tokyo, Japan), extension tube attached, that slipped into a Kodak Series VI lens hood (Eastman Kodak Company). The lens hood was attached to a 25-mm friction mount that fit snugly onto the microscope third ocular (photomicrographic) tube. This diameter was a longtime standard, although other diameters such as 30 mm are now in use. If all else fails, making an adapter coupler is a simple task for your local machinist, and I've made them out of wood using the appropriately sized Forstner bit.

If properly done, the microscope field of view will fill the camera's film frame. Focus is accomplished using the microscope focusing knobs. Routinely, I find it convenient to photograph a stage micrometer at the beginning of each roll using the objective lenses required for that project. This simplifies the making of scale bars later.

Making exposures. To use film, one needs an exposure meter, which is sometimes built into the camera. The proper exposure must be calibrated for photomicrography by running test rolls at different shutter speeds, and using the various objective lenses and filters as desired. At the outset you should run several test rolls of film. With each roll, keep a detailed exposure work sheet.

Film choice. In general, film speed matters little as silver grain size is not usually an issue at the enlargement factors used for publication figures. Faster films (EI 200–400) work well. They allow a faster shutter speed that can mitigate most vibration problems, but you should make an effort to minimize all recognizable sources of movement at the time of exposure (see "Mitigating Vibration Problems").

Even if you have calibrated the shutter speeds for your chosen film, you should "bracket" exposures. Make three shots per view: one a half-stop underexposed, one considered properly exposed, and one a half-stop overexposed. The best image contrast could be in any of the three.

Tissue Preparation for the SEM

Most of the numerous techniques for making observations using the scanning electron microscope (SEM) are beyond the scope of this guide. However, the preparation of plant tissues for use with the SEM can often be done alongside tissue preparations for light microscopy. A few suggestions are included here.

Surfaces of tissues, or cut surfaces, will often yield much information using the SEM. Fossil material or contemporary tissues can be observed from dried, restored, or fixed collections. Usually, wet or fixed plant tissue is dehydrated in a sequence of alcohols, then critical point dried, and sputter-coated with gold before placing under the electron beam.

Some contemporary makers of desktop instruments claim that gold-coating is unnecessary. However, in my experience, failing to coat is a bad idea. The absence of a metallic coating causes too much "charging," a build-up of electrons on the specimen. The resulting "white out" develops before you have finished navigating around the specimen, which makes it impossible to obtain a properly contrasted image.

Ethanolic Dehydration

Spellenberg and Bacon (1996) used FAA-fixed oak leaves, which were washed in 30% ethanol, dehydrated in absolute ethanol, critical point dried, and sputter-coated with gold. This is a commonly used sequence for plant tissues.

Methanol as Rapid Fixative

Neinhuis and Edelmann (1996) recommend the use of pure methanol treatment of tissues. Methanol instantly fixes cell walls that are elastically extended. Because of the speed of fixation, there is no specimen shrinkage. The authors hold that this preserves cell dimensions comparable with living conditions. This treatment works especially well for plant epidermal surfaces and other delicate specimens. In addition it renders a time-consuming fixation-dehydration sequence unnecessary.

Dimethoxypropane Dehydration

Maser and Trimble (1977) have recommended using 2,2-dimethoxypropane (DMP) as a one-step dehydrating agent. Muller and Jacks (1975) acidified the DMP by adding one drop of concentrated HCl per 100 ml of DMP. Specimens are fixed as usual, rinsed with water, drained, and immersed in the acidified DMP. Timing is a few minutes to overnight.

The presence of water in a specimen causes DMP to undergo hydrolysis in an endothermic reaction, becoming methanol and acetone. Results are reported to be just as good as moving the specimens through a graded series of alcohols. Tissues are then critical point dried and coated. DMP is a flammable solvent. See Johnson et al. (1976) for a discussion of DMP prior to SEM observation; and also Postek and Tucker (1976), who used acidified DMP prior to resin-embedment and transmission electron microscopy.

Silica Gel

(Keating). One often has a choice of viewing the surface of leaves from liquid-preserved specimens or when using the herbarium voucher of the same collection. For my material (*Salix* leaves), tests demonstrate that dried leaf surfaces produce detail that cannot be distinguished from that of specimens dehydrated from liquid storage.

Beginning with pink silica gel granules, place them in open Petri dishes in thin layers in a paraffin oven at the usual 60°C temperature.

After about 6 hours, the silica gel turns blue and can be stored in tight-lidded jars. In this dehydration process the silica gel loses about 40% of its (aqueous) weight.

Place the leaf pieces in a vial or empty film canister on top of a few layers of dehydrated silica gel and leave overnight. The next day the specimen can be attached to a stub, gold-coated, and observed using the SEM.

TABLE 1 Plant Tissue Dye Responses in Calcium Chloride Mounts[1]

Dye[2]	Lignin	Cellulose	Collenchyma	Metachromasy	Differentiation
Cresyl violet acetate (CVA)	Violet, blue	Yellow, tan	Red, tan	+++	+++
Thionin	Blue	Rose, tan	Reddish	+++	+++
Hematoxylin[3]	Red, orange	Tan, grey	Red tan	+++	+++
Iodine-potassium iodide	Yellow/ orange	Rose	Rose	+++	+++
Methylene blue[4]	Green, blue	None	None, pink	++	+++
Toluidine blue O	Blue	None	None	+	+++
Safranin O	Red	Yellow, tan	Yellow, tan	+	+++
Methyl green	Green	None, gray	None, gray	+	+++
Chlorazol black E (CBE)	None	Gray	Blue	+	++
Orange G	Orange	Tan	Tan	+	++
Fast green FCF	Green	None	None	–	+++
Schiff's reagent	Red pink	None	None	–	+++
Phloroglucinol	Red	None	None	–	+++
Trypan blue	None, blue	Blue	Blue	–	++
Erythrosin B	None	Pink	Pink	–	++
Malachite green	Blue-green	None	None	–	++
Celestin blue B	None	Violet blue	Violet blue	–	++
Aniline blue WS	Blue	None	None	–	++

TABLE 1 Plant Tissue Dye Responses in Calcium Chloride Mounts *(Cont.)*

Dye[2]	Lignin	Cellulose	Collenchyma	Metachromasy	Differentiation
Fluorescein	Yellow	Yellow	Yellow	–	+
Aniline blue black	None	Blue	Blue	–	+
Crystal violet	None, violet	Violet	Violet	–	+
Phloxine	None	Pink	Pink	–	–

Note: Triple plus means excellent result; fewer, less so. Minus means no response.

1 All dyes were made up in dilute 15% ethanolic solutions.

2 Additional dyes tested but not promising due to poor differentiation or blotchy results: alcian blue, alcian yellow GX, acid fuchsin, basic fuchsin, Bismarck brown Y, brilliant cresyl blue, Congo red, eosin B, lacmoid, solvent blue 38, superchrome B.

3 Striking but not permanent; dye may fade in 2–4 weeks.

4 Dye leaches into mountant; image not harmed.

Bibliography

Included at the end of this bibliography is a list of some useful specialist journals and electronic sources.

Abramowitz, M. 1987. Contrast Methods in Microscopy. Transmitted Light. Olympus, Lake Success, NY. 31 pp.

Alvin, K. L. & Boulter, M. C. 1974. A controlled method of comparative study for taxodiaceous leaf cuticles. Botanical Journal of the Linnean Society 69: 277–286. [Used chromium trioxide to clear interior tissues.]

Anderson, L. E. 1954. Hoyer's solution as a rapid permanent mounting medium for bryophytes. The Bryologist 57: 242–244.

Arnott, H. J. 1959. Leaf clearings. Turtox News 37: 192–194.

Artschwager, E. 1921. Use of chloriodide of zinc in plant histology. Botanical Gazette 71: 400.

Ayensu, E. S. 1967. Aerosol OT solution: an effective softener of herbarium specimens for anatomical study. Stain Technology 42: 155–156.

Baker, E. A. 1970. The morphology and composition of isolated plant cuticles. New Phytologist 69: 1053–1058.

Baum, S. F., Silk, W. K., et al. 1996. Effects of salinity on water use and xylem architecture in developing leaves of *Sorghum bicolor* (Poaceae). American Journal of Botany 83: 34 (abstracts).

Beacher, J. 2011. Microscope illumination: LEDs are the future. Microscopy Today July: 18–21. [A detailed comparison of conventional microscope lamps with light emitting diodes.]

Beerling, D. J. & Chaloner, W. G. 1992a. Stomatal density as an indicator of past atmospheric CO_2 concentration. Holocene 2: 71–78. [Used cellulose acetate impressions to assess stomatal density.]

Beerling, D. J. Chaloner, W. G, et al. 1992b. Variations in the stomatal density of *Salix herbacea* L. under the changing atmospheric CO_2 concentrations of late- and post-glacial time. Philosophical Transactions of the Royal Society B 336: 215–224. [Example of the acetone/cellulose acetate surface replica technique.]

Benninghoff, W. S. S. 1947. Use of trisodium phosphate with herbarium material and microfossils in peat. Science 108: 325.

Berlyn, G. P. & Miksche, J. P. 1976. Botanical Microtechnique and Cytochemistry. Iowa State University Press, Ames. 326 pp. [A thorough update of John Sass' well-known text by two of his students. Much added on cytochemistry and microscopy.]

Bersier, J. D. & Bocquet, G. 1960. Les méthodes d'éclairissment en vascularisation et en morphogénie végétales comparées. Archives des Sciences (Geneva) 13: 555–566.

Blaydes, G. W. 1939. The use of Bismarck Brown Y in some new staining schedules. Stain Technology 14: 105–110.

Bongers, F. & Popma, J. 1990. Leaf characteristics of the tropical rain forest flora of Los Tuxtlas, Mexico. Botanical Gazette 151: 354–365. [Techniques for leaf parameter descriptions and making stomatal prints using Xantopren Plus.]

Breil, D. 1965. Phloroglucinol and plant anatomy. Turtox News 43: 202–204.

Brewer, C. 1992. Responses by stomata on leaves to microenvironment conditions. pp. 67–75. In: Goldman, C. A. (editor). Proceedings of the 13th Workshop/Conference of the Association for Biology Laboratory Education (ABLE). Tested Studies for Laboratory Teaching, 191 pp. Association for Biology Laboratory Education. [An exercise on how stomata respond to environmental conditions. Includes information on making casts of leaf surfaces after different light treatments. www.zoo.utoronto.ca/able/volumes/vol-13/3-brewer/3-brewer.htm]

Bridson, D. & Forman, L. 1992. The Herbarium Handbook. Royal Botanic Garden, Kew, UK. [See also 1989 and 1998 editions. A general reference on herbarium and specimen management, with recipes for liquid preserved plant materials.]

Brown, B. V. 1993. A further chemical alternative to critical point drying for preparing small (or large) flies. Fly Times 11: 10.

Brown, C. A. 1960. Palynological Techniques. Privately published, Baton Rouge. 188 pp. [Contains numerous techniques for handling pollen and formulas for dyes and mountants.]

Brown, P. A. 1997. A review of techniques used in the preparation, curation and conservation of microscope slides at the Natural History Museum, London. The Biology Curator Issue 10 (Suppl). 34 pp. [A treasure trove of recipes, references, and discussion on this topic.]

Burrells, W. 1977. Microscope Technique. Fountain Press, London. 574 pp. [Many unusual details and arcane methods seldom found elsewhere.]

Bushnell, P. J. 2013. Solvents, ethanol, car crashes and tolerance. How risky is inhalation of organic solvents? American Scientist 101: 282–291.

Canright, J. E. 1953. The comparative morphology and relationships of the Magnoliaceae. II. Significance of the pollen. Phytomorphology 3: 355–365. [Dried pollen expansion method]

Carlquist, S. 1982. The use of ethylenediamine in softening hard plant structures for paraffin sectioning. Stain Technology 57: 311–317.

Chamberlain, C. J. 1932. Methods in Plant Histology. University of Chicago Press, Chicago. [A good source for many earlier techniques, it was the major source of its era.]

Cheng, V. C., Cheng, W. Y., et al. 2003. The study of epidermal replica by oblique illumination microscopy. Maize Genetics Cooperation Newsletter 77:28–29. http://mnl.maizegdb.org/mnl/77/54cheng.html. [This is the maize genetics and genomic database. Also, cf. Microscopy and Microanalysis 8: 1068–1069.]

Chiovetti, R. 2001. Another tip on cleaning microscope objectives. Microscopy Today 01–8: 36. [Lists solvents and practices to avoid when cleaning lens surfaces.]

Christophel, D. C., Kerrigan, R., et al. 1996. The use of cuticular features in the taxonomy of the Lauraceae. Annals of the Missouri Botanical Garden 83: 419–432. [Cuticles are removed for study using hydrogen peroxide.]

Clarke, T. 2012. Resolving low contrast microstructure using transmitted and reflected circular oblique illumination (COL). www.microscopy-uk.org.uk/mag/index.mag.html. [A detailed description of microscope modifications for this type of illumination.]

Coelho, V. P. M., Leite, J. P. V., et al. 2012. Anatomy, histochemistry, and phytochemical profile of leaf and stem bark of *Bathysa cuspidata* (Rubiaceae). Australian Journal of Botany 60: 49–60.

Conn, H. J. & Darrow, M. A, et al. (editors). 1960. Staining Procedures Used by the Biological Stain Commission. Ed. 2. Williams & Wilkins, Baltimore. 289 pp. [A comb-bound collection of recipes and techniques for plant and animal specimens. Paraffin technique is assumed but many fixatives and dyes also work well with hand techniques. Never reprinted or updated, it is still available as a PDF. See also Lillie (1969).]

Cougar, G. 2005. Comments on the use of plane irons and wood chisels for sectioning woody plant stems. Netnotes, Microscopy Today March: 58.

Cranwell, L. M. 1953. New Zealand pollen studies: the Monocotyledons: a comparative account. Auckland Institute and Museum Bulletin 3: 1–99. Harvard University Press, Cambridge, MA. [Lactic acid-based mountant.]

Cutler, D. F. 1969. Anatomy of the Monocotyledons. IV Juncales. Oxford University Press, New York. 357 pp. [Scraping technique for examining leaf surfaces.]

Dacar, M. A. & Giannoni, S. M. 2001. Technical note: a simple method for preparing reference slides of seed. Journal of Range Management 54: 191–193. [Sodium bicarbonate macerating method.]

Dahl, A. O. 1952. The comparative morphology of the Icacinaceae. VI. The pollen. Journal of the Arnold Arboretum 33: 252–286. [Lactic acid-based mountant.]

Darrow, M. A. 1940. A simple staining method for histology and cytology. Stain Technology 15: 67–68.

De Fossard, R. A. 1969. Development and histochemistry of the endothecium in the anthers of in vitro grown *Chenopodium rubrum* L. Botanical Gazette 130: 10–22.

Dizeo de Strittmatter, C. G. 1980. Coloración con Violeta de Cresyl. Boletín de la Sociedad Argentina de Botanica 29: 273–276.

Dolph, G. E. 1977. The effect of different calculational techniques on the estimation of leaf area and the construction of leaf size distributions. Bulletin of the Torrey Botanical Club 104: 264–269. [A review of techniques for estimating leaf areas, with the recommendation to use a dot planimeter for greatest accuracy.]

Duke, C. V. 2004. Lens cleaning: best practice review. Microscopy Today March/April 2004: 42, 44.

Eckardt, N. 2002. Probing the mysteries of lignin biosynthesis: the crystal structure of caffeic acid/5-hydroxyferulic acid 3/5-0-methyltransferase provides new insights. The Plant Cell 14: 1185–1189.

Ejima, H., Richardson, J. J., et al. 2013. One-step assembly of coordination complexes for versatile film and particle engineering. Science 341: 154–157. [A specimen coating technique using tannic acid and ferric chloride.]

Ellis, B., Daly, D. C., et al. 2009. Manual of Leaf Architecture. The New York Botanical Garden Press and Comstock Publishing Associates, Ithaca. 190 pp. [The most recent revision of the work originated by L. J. Hickey (1973). Beyond being a thorough illustrated glossary, it furnishes the only organized, descriptive protocol for describing dicot leaves. Useful for some monocot groups also.]

Erdtman, G. 1952. Pollen Morphology and Plant Taxonomy. Angiosperms. Almquist & Wiksell, Stockholm; Chronica Botanica, Waltham, Mass. [This remains an excellent introduction to pollen morphology. His terminology differs somewhat from that of Faegri and Iverson (below).]

Evered, C. 2009. Specimen preparation, in: Netnotes. Microscopy Today, 17(1): 59. [Toluidine blue use, especially for resin-embedded specimens.]

Faegri, K. & Iversen, J. 1989. Textbook of Pollen Analysis. Ed. 4. John Wiley & Sons, Chichester, UK. [An excellent guide to this field.]

Feder, N. & O'Brien, T. P. 1968. Plant microtechnique: some principles and new methods. American Journal of Botany 55: 123–142. [A review and critique of traditional microtechnique with some new ideas.]

Foster, A. S. 1949. Practical Plant Anatomy. Ed. 2. D. Van Nostrand. New York. 228 pp. [This is both a textbook and laboratory guide with study examples. The appendix has good suggestions for hand-sectioning and basic technique suitable for the student laboratory as well as research.]

Foster, B. 2011. Light-emitting diodes: a new solution for fluorescence. American Laboratory Oct: 36–38. [A review of LED advances with tables that include excitation wavelengths and their uses.]

Fox, C. H., Johnson, F. B., et al. 1985. Formaldehyde fixation. Journal of Histochemistry and Cytochemistry 33: 845–853. [Excellent review of its history and chemistry.]

Fuchs, C. 1963. Fuchsin staining with NaOH clearing for lignified elements of whole plants or plant organs. Stain Technology 38: 141–144.

Gahan, P. B. 1984. Plant Histochemistry and Cytochemistry. An Introduction. Academic Press, New York. [A detailed review of dyes and how they work on the various types of plant tissues.]

Garay, L. A. 1979. Systematics of the genus *Stelis* Sw. Botanical Museum Leaflets Harvard University 27: 167–259. [Ammonia method to restore dried orchid flowers.]

Gardner, R. O. 1975. An overview of botanical clearing technique. Stain Technology 50: 99–105. [Good discussion of the various approaches and chemical effects of each. The bibliography lists many well-known methods.]

Gatenby, J. B. & Beams, H. W. (editors). 1950. The Microtomist's Vade-mecum. A Handbook of Methods of Animal and Plant Microscopic Technique. Blakiston. Philadelphia.

Gerlach, D. 1969. Botanische Mikrotechnik. Stuttgart. Georg Thieme Verlag. [Acetocarmine use for pollen staining.]

Gitz, D. C. & Baker, J. T. 2009. Methods for creating stomatal impressions directly onto archivable slides. Agronomy Journal 101: 232–236. [A general review, plus new method: thick plastic sheets are used to make impressions that serve directly as slides.]

Gloser, J. 1967. Some problems of the determination of stomatal aperture by the microrelief method. Biologia Plantarum (Prague) 9: 28–33. [A replica method.]

Gray, P. 1954. The Microtomist's Formulary and Guide. New York. Blakiston.

Gray, P. 1958. Handbook of Basic Microtechnique. Ed. 2. McGraw-Hill, New York.

Groot, J. 1969. The use of silicone rubber plastic for replicating leaf surfaces. Acta Botanica Neerlandica 18: 703–708.

Hayat, M. A. 2000. Care and use of microwave oven for diagnostic pathology. Microscopy Today 00–3: 29.

Haynes, W. M. (editor). 2010. CRC Handbook of Chemistry and Physics. CRC Press, Boca Raton. 91st edition. [The "go to" reference for all things chemical. Previous editions of the past 20 years are quite adequate for basic chemical data relevant to techniques covered in this guide.]

Herr, J. M. Jr. 1971. A new clearing-squash technique for the study of ovule development in angiosperms. American Journal of Botany 58: 785–790. [The introduction of "Herr's 4½" clearing method.]

Herr, J. M. Jr. 1972a. Applications of a new clearing technique for the investigation of vascular plant morphology. Journal of the Elisha Mitchell Scientific Society 88: 137–143.

Herr, J. M. Jr. 1972b. An extended investigation of the megagametophyte in *Oxalis corniculata* L. Advances in Plant Morphology (1972): 92–101.

Herr, J. M. Jr. 1973a. The use of Nomarski Interference Microscopy for the study of structural features in cleared ovules. Acta Botanica Indica 1: 35–40.

Herr, J. M. Jr. 1973b. The cytological effects of several fixatives on the ovules of *Cassia abbreviata* Oliver var. *granitica* Bak. New Botanist 1: 118–126.

Herr, J. M. Jr. 1974. A clearing-squash technique for the study of ovule and megagametophyte development in angiosperms. pp. 230–235. In: Radford, A. E., Dickison, W. C., Massey, J. R., Bell, C. R. Vascular Plant Systematics. Harper & Row, New York. [Explained in detail here and with all Herr's other refs cited.]

Herr, J. M. Jr. 1982. An analysis of methods for permanently mounting ovules clearing in four-and-a-half type clearing fluids. Stain Techology 57: 161–169.

Herr, J. M. Jr. 1992. New uses of calcium chloride solution as a mounting medium. Biotechnic & Histochemistry 67: 9–13. [The reintroduction of a slide mountant that was well known in the mid-nineteenth century.]

Herr, J. M. Jr. 1993. Clearing techniques for the study of vascular plant tissues in whole structures and thick sections, Chapter 5, pp. 63–84. In: Tested Studies for Laboratory Teaching. Vol. 5. Goldman, C. A., Hauta, P. L., et al. (editors). Proceedings of the 5th Workshop Conference of Association for Biology Laboratory Education (ABLE). 115 pp. [A detailed description of the technique especially for students, and a listing of Herr's related publications.] [www.zoo. utoronto.ca/able]

Hesse, M., Habritter, H., et al. 2009. Pollen Terminology. An Illustrated Handbook. Springer, New York. 261 pp. [Covers all aspects of pollen morphology and analysis, mostly using examples from the SEM, and with illustrations that cannot be improved upon.]

Hewlett, B. R. 2002. Penetration rates of formaldehyde. Microscopy Today 10: 30.

Hickey, L. J. 1973. Classification of the architecture of dicotyledonous leaves. American Journal of Botany 60: 17–33. [The first thorough attempt in a century to systematize leaf feature description. See the volume by Ellis et al. (2009) for the current permutation.]

Hickey, L. J. 1979. A revised classification of the architecture of dicotyledonous leaves. In: Metcalfe, C. R., Chalk, L. (editors). Anatomy of the Dicotyledons. Ed. 2. Vol. 1. pp. 25–39. Oxford: Clarendon Press. [A revised version of Hickey's (1973) system.]

Horobin, R. & Kiernan, J. (editors). 2002. Conn's Biological Stains: A Handbook of Dyes, Stains and Fluorochromes for Use in Biology and Medicine. Taylor & Francis, London and New York. 502 pp. [Same as 10th edition, Bios Scientific Publishers, Oxford, UK. First edition with this publisher. This book and its predecessors (see Lillie below) has a 75-year history, with editions 8 and 9 edited by R. D. Lillie. It is the only such encyclopedia covering structural formulas, spectra, chemistry, nomenclature, and uses.]

Ikuse, M. 1956. Pollen grains of Japan. Kirokawa Publishing, Tokyo. 300 pp. [Resin-mounted pollen.]

Ilarslan, H., Palmer, R. G., et al. 1997. Quantitative determination of calcium oxalate and oxalate in developing seeds of soybean (Leguminosae). American Journal of Botany 84: 1042–1046. [Descriptions of several methods of oxalate detection including urinalysis diagnosis kits.]

Iwanowska, A., Tykarska, T., et al. 1994. Localization of phenolic compounds in the covering tissues of the embryo of *Brassica*. Annals of Botany 74: 313–320.

James, P. 2012. www.microscopy-uk.org.uk/mag/index.mag.html. [Four detailed articles on circular oblique illumination (COL).]

Jansen, S., Kitin, P., et al. 1998. Preparation of wood specimens for transmitted light microscopy and scanning electron microscopy. Belgian Journal of Botany 131: 41–49. [Glycerin was used to soften wood for sectioning.]

Jensen, W. A. 1962. Botanical Histochemistry. Principles and Practice. W. H. Freeman, San Francisco. 408 pp. [Procedures are carefully explained but require a well-equipped laboratory.]

Johansen, D. A. 1940. Plant Microtechnique. McGraw-Hill. New York. 523 pp. [On this topic, this text probably had the longest print run. It is considered the "go to" book by several generations of plant anatomists. It contains numerous techniques plus commentary on all those included.]

Johnson, W. S., Hooper, G. R., et al. 1976. 2, 2-dimethoxypropane, a rapid dehydrating agent for scanning electron microscopy. Micron 7: 305–306.

Joubert, J. & Sharma, D. 2011. Using CMOS cameras for light microscopy. Microscopy Today July: 22–26. [A favorable comparison of CMOS sensors vs. CCD sensors for digital photomicrography.]

Kappe, C. O. 2001. Speeding up solid-phase chemistry by microwave radiation: a tool for high-throughput synthesis. American Laboratory (May 2001): 13.

Kappelle, M. & Leal, M. E. 1996. Changes in leaf morphology and foliar nutrient status along a successional gradient in a Costa Rican upper montane *Quercus* forest. Biotropica 28: 331–344. [Used a polar planimeter for measuring leaf area, plus a technique for deriving specific leaf weight.]

Karabourniotis, G., Tzobanoglou, D., et al. 2001. Epicuticular phenolics over guard cells: exploitation for *in situ* stomatal counting by fluorescence microscopy and combined image analysis. Annals of Botany 87: 631–639.

[This paper exploits differences in fluorescence emissions between guard cell phenolics and neighboring epidermal cells for making stomatal counts. Useful bibliography on stomatal studies, fluorescence, and image analysis.]

Keating, R. C. 1996. Anther investigations: a review of methods, pp. 255–271. In: D'Arcy, W. G. & Keating, R. C. (editors). The Anther: Form, Function and Phylogeny. Cambridge University Press, New York. 336 pp. [A review of general microtechnique with emphasis on flower and anther studies. Dissection, hand-sectioning, and wet mounts are discussed, and, briefly, pollen methods.]

Keating, R. C. 2000. Anatomy of the young vegetative shoot of *Takhtajania perrieri* (Winteraceae). Annals of the Missouri Botanical Garden 87: 335–346. [Methods section described the iodine-calcium chloride method of clearing.]

Kiernan, J. A. 1990. Histological and Histochemical Methods. Theory and Practice. Ed. 2. Oxford: Pergamon Press. [A comprehensive treatment of this topic.]

Kiernan, J. A. 1997. Making and using aqueous mounting media, or why buy a premixed medium when it's cheap and easy to make your own? Microscopy Today 97-10: 16–17. [Five recipes of different types.]

Kiernan, J. A. 1999. Long-lasting aqueous mountants for light microscopy. Microscopy Today 99-1: 3. [Comments on several mounts including fructose syrup.]

Kisser, J. 1935. Bemerkungen zum Einschluss in Glyzerin-Gelatine. Zeitschrift für wissenschaftliche Mikroskopie und für microskopische Technik 51: 372–374.

Kong, H. Z. 2001. Comparative morphology of leaf epidermis in the Chloranthaceae. Botanical Journal of the Linnean Society 136: 279–294. [A technique for separating leaf epidermal layers using macerating fluid.]

Kubinova, L. 1994. Recent stereological methods for measuring leaf anatomical characteristics: estimation of the number and sizes of stomata and mesophyll cells. Journal of Experimental Botany 45: 119–127. [Discussion of proper random sampling methods for leaf surface features.]

Kuchachka, B. F. 1977. Sectioning refractory woods for anatomical studies. USDA Forest Service Research Note FPL-0236: 1–9. [See also Microscopica Acta 80: 301–307, 1978.]

Kutscha, N. & Gray, J. 1972. The suitability of certain stains for studying lignification in balsam fir (*Abies balsamea* L.) Mill. Life Sciences and Agricultural

Experiment Station. Technical Bulletin 53: 70 pp. [A detailed review of experiments with numerous dyes in the staining of wood.]

Lersten, N. R. 1967. An annotated bibliography of botanical clearing methods. Iowa State Journal of Science 41: 481–486. [A thorough listing giving some history and nature of the methods cited. Annotations indicate ingredients of the procedures.]

Lillie, R. D., et al. (editors). 1969. H. J. Conn's Biological Stains. Ed. 8 [up through ed. 9]. Lippincott Williams & Wilkins, Baltimore. 498 pp. [Aptly subtitled "A Handbook on the Nature and Uses of the Dyes Employed in the Biological Laboratory." See also Conn et al. (1960) and Horobin & Kiernan (2002).]

Maácz, C. J. & Vágás, E. 1961. A new method for staining of cellulose and lignified cell walls. Microskopie 16: 40–43. [The use of Astra blue, auramine, and safranin.]

Martin, G., Myres, D. A., et al. 1991. Characterization of plant epidermal lens effects by a surface replica technique. Journal of Experimental Botany 42: 581–587.

Marx, A. & Sachs, T. 1977. The determination of stomata pattern and frequency in *Anagallis*. Botanical Gazette 138: 385–392.

Maser, M. D. & Trimble, J. J. III. 1977. Rapid chemical dehydration of biologic samples for scanning electron microscopy using 2, 2-dimethoxypropane. Journal of Histochemistry & Cytochemistry 25: 247–251.

Maurizio, A. 1953. Report of the IUBS International Commission for Bee Biology 1952. Bee World 34: 48–51. [Pollen lipid removal and glycerine jelly mountant.]

Mbagwu, F. N. & Edeoga, H. O. 2005. Morphology of the leaf epidermis in some *Vigna* species. [From manuscript of unknown final publication.]

McAuliffe, G. 2009. Specimen preparation, in: Netnotes. Microscopy Today 17: 59. [Toluidine blue use, especially for resin-embedded specimens.]

McCrone, W. C. 1999. Another way to prepare permanent mounts for light microscopy. Microscopy Today 99: 6.

Miller, N. A. & Ashby, W. C. 1968. Studying stomates with polish. Turtox News 46: 322–324.

Miller, R. H. 1982. Apple fruit cuticles and occurrence of pores and transcuticular canals. Annals of Botany 50: 355–371. [Cuticle separation using on zinc chloride and HCl.]

Miller, R. H. 1986. The prevalence of pores and canals in leaf cuticular membranes. 2. Supplemental studies. Annals of Botany 57: 419–434.

Morawetz, J. J. 2013. A clearing protocol for whole tissues: an example using haustoria of Orobanchaceae. Applications in Plant Sciences 1: 1200361 (3 pp.). [The use of Stockwell's bleach followed by lactic acid and chloral hydrate.]

Morley, T. 1949. Staining of plant materials in NaOH. Stain Technology 24: 231–235. [Includes a list of 20 different stains and their results for leaves cleared in NaOH.]

Moss, A. G. 1999. Air tables are not always needed for vibration isolation. Microscopy Today 99: 27. [Methods of dampening building vibration for higher-resolution microscopy.]

Muller, L. L. & Jacks, T. J. 1975. Rapid chemical dehydration of samples for electron microscopic examinations. Journal of Histochemistry & Cytochemistry 23: 107–110. [The use of acidified DMP.]

Neinhuis, C. & Edelmann, H. G. 1996. Methanol as a rapid fixative for the investigation of plant surfaces by SEM. Journal of Microscopy 184: 14–16. [For SEM images, anhydrous methanol immersion produced better results than FAA-fixed plant materials.]

Netnotes column. 2005. Microtomy–sectioning woody plant stems. Microscopy Today. [Multiple authors responded to this topic.]

O'Brien, T. P. & McCully, M. E. 1981. The study of Plant Structure. Principles and Selected Methods. Termarcarphi Pty. Ltd., Melbourne. [Excellent general microtechnique text that includes explanations and theory.]

O'Brien, T. P. & von Teichmann, I. 1974. Autoclaving as an aid in the clearing of plant specimens. Stain Technology 49: 175–176. [Specimens are autoclaved in capped bottles, in various solutions depending on difficulty of clearing.]

Ogburn, R. M. & Edwards, E. J. 2009. Anatomical variation in Cactaceae and relatives: trait lability and evolutionary innovation. American Journal of Botany 96: 391–408. [A demonstration of Keating's calcium chloride/cresyl violet acetate method, complete with color cover photo.]

Ogden, E. C., Raynor, G. S, et al. 1974. Manual for Sampling Air-borne Pollen. Hafner Press, New York [Source of Calberla's mountant formula, and much else.]

Olowokudejo, J. D. 1990. Comparative morphology of leaf epidermis in the genus *Annona* (Annonaceae) in West Africa. Phytomorphology 40: 407–422.

Parr, J. F. 2002. A comparison of heavy liquid flotation and microwave digestion techniques for the extraction of fossil phytoliths from sediments. Review of Palaeobotany and Palynology 120: 315–336. [A review of techniques.]

Parr, J. F., Dolic, V., et al. 2001. A microwave digestion method for the extraction of phytoliths from herbarium specimens. Review of Palaeobotany and Palynology 116: 203–212. [A complex process using pressure vessels.]

Pearsall, D. M. 1989. Paleoethnobotany. Academic Press, New York. [Processing of vegetative samples in archeological research. Flotation and identification of plant remains from field sites.]

Peterson, R. L., Peterson, C. A., et al. 2008. Teaching Plant Anatomy Through Creative Laboratory Exercises. National Research Council Press, Ottawa. 254 pp. plus CD. [A course book that includes methods of hand-sectioning, clearing, macerations, plus recipes for stains and other reagents.]

Phillips, T. E. (editor). 2004. Netnotes. Microscopy Today Nov: 56.

Piekos, W. B. 1999. Diffracted-light contrast enhancement: a reexamination of oblique illumination. Microscopy Research and Technique 46(4/5): 334–337. [As reviewed in Literature Highlights, Microscopy and Analysis 1999, Nov: 33.]

Piekos, W. B. 2003. Apparatus and method for producing diffracted light contrast enhancement in microscopes. US Patent 6,600,598. [A detailed description with ample illustrations of the DLC set-up.]

Piekos, W. B. 2006. Diffracted light contrast: improving the resolution of a basic light microscope by an order of magnitude. Microscopy Today Nov: 10–15. [A brief, illustrated review of the principles given in the 1999 paper.]

Piperno, D. R. 1988. Phytolith Analysis. An Archeological and Geological Perspective. Academic Press, San Diego. 280 pp. [Plant minerals, silica bodies, and their study by archeobotanists. Such materials are all that remain in many sites where carbon-based carcasses have decomposed.]

Piperno, D. R. 2006. Phytoliths: A Comprehensive Guide for Archeologists and Paleoecologists. Altanira Press, Lanham, MD.

Postek, M. T & Tucker, S. C. 1976. A new short chemical dehydration method for light microscopy preparations of plant material. Canadian Journal of Botany 54: 872–875. [The use of dimethoxypropane (DMP) for specimen dehydration.]

Purvis, M. J., Collier, D. C., et al. 1966. Laboratory Techniques in Botany. Ed. 2. London. Butterworth. 439 pp. [A broadly based text including, in addition to anatomical plant microtechnique, methods in physiology, culture techniques, soil analysis, and much else.]

Radford, A. E., Dickison, W. C., et al. 1974. Vascular Plant Systematics. Harper & Row, New York. 891 pp. [An encyclopedia/dictionary/techniques compendium of all things systematic.]

Rao, C. K. 1977. A technique for the revival of herbarium specimens for floral dissections and anatomical studies. Current Science 46: 720. [Used EDTA, glycerol, acetic acid mixture.]

Rao, T. A. 1957. Comparative morphology and ontogeny of foliar sclereids in seed plants. I. *Memecylon*. Phytomorphology 7: 306–330.

Rawlins, T. E. 1933. Phytopathological and Botanical Research Methods. John Wiley & Sons., New York.

Reid, P. D, Pont-Lezica, R. F., et al. (editors). 1992. Tissue Printing: Tools for the Study of Anatomy, Histochemistry, and Gene Expression. Academic Press, New York. 188 pp. [A comb-bound collection of techniques inspired by the work of Joseph E. Varner. Freshly sliced plant organs are pressed or printed onto nitrocellulose or other membranes. The organ's structure or its chemistry can then be observed from these replicas.]

Rodin, R. J. & Davis, R. E. 1967. The use of papain in clearing plant tissues for whole mounts. Stain Technology 42: 203–206.

Rogers, L. A., Dubos, C., et al. 2005. Comparison of lignin deposition in three ectopic lignification mutants. New Phytologist 168: 123–140.

Rollins, R. C. 1955. The Archer method for mounting herbarium specimens. Rhodora 57: 294–299. [Contains the formula for a "glue" used for many years in large herbaria for mounting plant specimens. It contains a styrene resin, ethyl cellulose, toluene, and methanol.]

Romberger, J. A. 1966. Microdissecting tools made from microminiature drills and hypodermic needles. Bioscience May 1966: 373–374.

Rost, F. & Oldfield, R. 2000. Photography through the Microscope. Cambridge University Press, New York. 278 pp. [Well-written and illustrated introduction to film and early digital photomicrography. Covers most general issues.]

Rudall, P. J. & Clark, L. 1992. The megagametophyte in Labiatae. pp. 65–84, In: Harley, R. M. & Reynolds, T. (editors). Advances in Labiatae Science. Royal Botanic Gardens, Kew. [Used modified Herr's clearing fluid for ovule studies.]

Runions, C. J. 1998. Surface replication. Microscopy Today 98-8: 28. [Use of ignition sealant.]

Ruzin, S. E. 1999. Plant Microtechnique and Microscopy. Oxford University Press, New York. 322 pp. [A thorough update on these topics together with rationales for the techniques.]

Sampson, J. 1961. A method for replicating dry and moist surfaces for examination by light microscopy. Nature 191: 932–933. [Silicone rubber.]

Sandoval, Z. E. & Rojas Leal, A. 2005. Técnicas aplicadas al estudio de la anatomiá vegetal. National Autonomous University of Mexico, Mexico City. [Good introduction to plant structural studies that covers topics from proper dissection through embedding and sectioning. Staining techniques for various tissues. Written and published in Spanish]

Sass, J. E. 1958. Botanical Microtechnique. Ed. 3. Iowa State University Press, Ames. 228 pp. [With printings through 1968. General microtechnique is well-explained. It includes commentary on plant organs and procedures.]

Schaede, R. 1940. Über den Feinbau von Parenchymmembranen. Berichte Deutschen Botanischen Gesellschaft 58: 275–290.

Schmid, R. 1977. Stockwell's bleach, an effective remover of tannin from plant tissues. Botanische Jahrbucher für Systematik 98: 278–287.

Schmid, R. 1981. Polyethylene bags for permanently sealing jars of pickled specimens. Taxon 30: 369–370.

Schmid, R. 1982. Sonication and other improvements on Jeffrey's technique for macerating wood. Stain Techology 57: 293–299.

Schmid, R. & Turner, M. D. 1977. Contrad 70, an effective softener of herbarium material for anatomical study. Taxon 26: 551–552.

Schnichnes, D., Nemson, J. A., et al. 1999. Microwave protocols for paraffin microtechnique and in situ localization in plants. Microscopy and Microanalysis 4: 491–496.

Seago, J. L. Jr., Peterson, C. A., et al. 1999. Development of the endodermis and hypodermis of *Typha glauca* Godr. and *Typha angustifolia* L. roots. Canadian Journal of Botany 77: 122–134.

Shobe, W. R. & Lersten, N. R. 1967. A technique for clearing and staining gymnosperm leaves. Botanical Gazette 128:150–152. [Clearings using sodium hydroxide and bleach.]

Silverman, J. 1999. D-Limonene: a serviceable and safe routine clearing agent. Microscopy Today 99-10: 18.

Simpson, J. L. S. 1929. A short method of clearing plant tissues for anatomical studies. Stain Technology 4: 131–132.

Sinclair, C. B. & Dunn, D. B. 1961. Surface printing of plant leaves for phylogenetic studies. Stain Technology 36: 299–304. [A technique that adapts Archer's medium, a once-popular herbarium specimen mounting glue.]

Slater, E. M. & Slater, H. S. 1992. Light and Electron Microscopy. Cambridge University Press, New York. 312 pp. [A grounding in the physics of optics with mathematical formulas. Excellent diagrams.]

Smith, B. B. 1973. The use of a new clearing technique for the study of early ovule development, megasporogenesis, and megagametogenesis in five species of *Cornus* L. American Journal of Botany 60: 322–338. [Uses Herr's clearing fluids.]

Smith, R. F. 1990. Microscopy and Photomicrography. A Working Manual. CRC Press, Boca Raton. 135 pp. [Good introduction to microscopy and film-based photomicrography, but with useful general tips.]

Spellenberg, R. & Bacon, J. R. 1996. Taxonomy and distribution of a natural group of black oaks of Mexico (Quercus, section Lobatae, subsection Racemiflorae). Systematic Botany 21: 85–99. [A technique for preparing specimens for SEM observations.]

Stebbins, G. L. Jr. 1938. A bleaching and clearing method for plant tissues. Science: 87: 21–22.

Stern, W. L. 2004. Letter to editor regarding fixation. Plant Science Bulletin 50: 50. [Recommends against using "Kew Cocktail" as it turns liquid preserved organs to "mush." He argues it is better to use one based on ethanol, formalin, and acetic acid.]

Stern, W. L., Curry, K. J., et al. 1986. Staining fragrance glands in orchid flowers. Bulletin of the Torrey Botanical Club 113: 288–297.

Sumner, A. J. & Sumner, B. E. H. 1969. A Laboratory Manual of Microtechnique and Histochemistry. Blackwell, Oxford.

Sundberg, M. D. & Zahn, S. G. 1985. A microscopic technique to measure mesophyll succulence. American Journal of Botany 72: 1654–1656. [Authors used an ocular grid to count mesophyll features.]

Tanaka, N. 1940. Chromosome studies in Cyperaceae. VI: Pollen development and additional evidence for the compound chromosomes in *Scirpus lacustris* L. Cytologia 10: 348–352.

Tillson, A. H. & Bamford, R. 1938. The floral anatomy of the Aurantioideae. American Journal of Botany 25: 780–793.

Titford, M. 2002. Save that dye! Microscopy Today Sept/Oct: 31.

Toscano de Brito, A. L. V. 1996. The use of concentrated ammonia as an excellent medium for the restoration of orchid pollinaria: an example from the subtribe Ornithocephalinae (Orchidaceae). Lindleyana 11: 205–210.

Traverse, A. 2008. Paleopalynology. Ed. 2. Springer, New York. 813 pp. [Encyclopedic source on techniques for pollen study with emphasis on the fossil record.]

Valdés-Reyna, J. & Hatch, S. L. 1995. Anatomical study of *Erioneuron* and *Dasychloa* (Poaceae: Chloridoideae: Eragrostideae) in North America. Sida 16: 413–426. [Photoflo for restoring dried specimens.]

Valkama, E., Koricheva, J., et al. 2005. Delayed induced responses of birch glandular trichomes and leaf surface lipophilic compounds to mechanical defoliation and simulated winter browsing. Oecologia 146: 385–393. [Leaf surface impressions were made using superglue.]

Valkama, E., Salminen, J. P., et al. 2003. Comparative analysis of leaf trichome structure and composition of epicuticular flavonoids in Finnish birch species. Annals of Botany 91: 643–655. [Surface impressions were made on a slide using glue.]

Van Cleave, H. J. & Ross, J. A. 1947. Use of trisodium phosphate in microscopical technic. Science 106: 194.

Vega, A. S., Castro, M. A., et al. 2008. Anatomy and histochemical localization of lipid secretions in Brazilian species of *Panicum* sect. *Lorea* (Poaceae, Panicoideae, Paniceae). Annals of the Missouri Botanical Garden 95: 511–519. [Softening method using hydrofluoric acid.]

Venning, F. 1954. Handbook of Advanced Plant Microtechnique. W. C. Brown, Dubuque. 96 pp. [A collection of published techniques and schedules, mostly focused on embedding and sectioning. Some whole mount and simpler techniques are also included.]

Vierheilig, H., Coughlan, A. P., et al. 1998. Ink and vinegar, a simple staining technique for arbuscular-mycorrhizal fungi. Applied and Environmental Microbiology 64: 5004–5007.

Villani, T. S., Korach, A. R, et al. 2013. An improved clearing and mounting solution to replace chloral hydrate in microscopic applications. Applications in Plant Sciences 5: 1300016. [Introduced Visikol.]

Vogel, S. & Cocucci, A. 1988. Pollen threads in *Impatiens*: their nature and function. Beiträge zur Biologie der Pflanzen 63: 217–287. [An example of use of cellulose staining as described in Maácz & Vágás (1961)]

Webb, L. J. 1959. A physiognomic classification of Australian rain forests. Journal of Ecology 47: 551–570. [The Raunkiaer system of leaf size classification.]

Weyers, J. D. B. & Johansen, L. G. 1985. Accurate estimation of stomatal aperture from silicone rubber impressions. New Phytologist 101: 109–115.

Weyers, J. D. B. & Travis, A. J. 1981. Selection and preparation of leaf epidermis for experiments in stomatal physiology. Journal of Experimental Botany 32: 837–850. [Used surface replicas in studies of living leaves.]

Whetten, R. & Sederoff, R. 1995. Lignin biosynthesis. The Plant Cell 7: 1001–1013.

Willey, R. L. 1971. Microtechniques: A Laboratory Guide. Macmillan, New York. 99 pp. [Focused on paraffin embedding and microtoming. Much general information on microscopy and slide analysis.]

Wilson, T. K. 1956. A simplified technique for securing wood samples. Journal of Forestry 54: 1. [Use of increment bored samples to make wood sections.]

Wilson, T. K & Shutts, C. F. 1957. A rapid wood maceration schedule using iron alum hematoxylin. Stain Technology 32: 149.

Wodehouse, R. P. 1935. Pollen Grains. Their Structure, Identification, and Significance in Science and Medicine. McGraw-Hill, New York.

Woodland, D. W. 2000 [and previous editions]. Contemporary Plant Systematics. Ed. 3. Andrews University Press, Burien Springs, MI. 569 pp. [Republished the Pohl solution, or Pohlstoffe, formula.]

Yasue, T. 1969. Histochemical identification of calcium oxalate. Acta Histochemistry & Cytochemistry 2: 83–95. [Review of methods to identify these crystals especially as found in kidney tissue.]

Zelitch, I. (editor). 1963. Stomata and water relations in plants. The Connecticut Agriculture Experiment Station, New Haven. Bulletin 664, 116 pp. [Early description of the silicone rubber positive/negative leaf casting technique.]

Zuloaga, F. O. & Morrone, O. 1996. Revisión de las especies Americanas de *Panicum* subgénero *Panicum* seccion *Panicum* (Poaceae) Panicoideae: Paniceae. Annals of the Missouri Botanical Garden 83: 200–280. [Prepared grass leaf sections following HF treatment.]

Journals

American Biotechnology Laboratory [Distributed without cost to professionals, it stresses biochemistry and separation science. Occasional good articles on germane topics such as polymerization, as well as materials of value to users of this guide.]

American Laboratory [Distributed without cost to professionals. Articles on all sorts of topics occasionally germane to microscopy and plant processing. The cover art often features inspiring LM photomicrographs.]

Biotechnic-Histochemistry (1991 to present) [This is a name change from Stain Technology (1926–1990). Both titles are good sources for techniques and stain applications. I especially value the older numbers of Stain Techology for simple, but still useful, protocols.]

Microscopy and Analysis [A British-based international journal with "The Americas" edition. Available by subscription, or free to professionals. It contains articles on all types of microscopy and interesting book reviews. See www. microscopy-analysis.com]

Microscopy Today [Published by Microscopy Society of America and distributed free to professionals. A good source of state-of-the-art machinery and techniques as well as occasional reviews of optical history. It has columns called *Microscopy 101* and *Net Notes* where users get help with specific problems from the community of microscope and microtechnique practitioners. Earlier volumes are especially strong in LM. See www. microscopy-today.com.]

Electronic Sources

Website citations are limited due to their constant flux.

www.biologicalstaincommission.org [Biological Stain Commission site.]

www.kenrockwell.com [A photographer's site where digital cameras and techniques are reviewed.]

www.microscopy-today.com [Microscopy Society of America site.]

www.microscopy.com [A free listserver for "NetNotes," a column in Microscopy Today. It has a Q&A format that covers mostly EM and other advanced techniques. It has some LM questions and techniques.]

www.microscopyu.com [Nikon site, covers many aspects of microscopy.]

www.olympusmicro.com [Olympus site, covers many aspects of microscopy.]

Search on: *light microscopy* [for numerous primers on the capability of the microscope.]

Search on: *digital photography* [for much information on the principles of this fast-changing new medium.]

Chemical and Formula Index

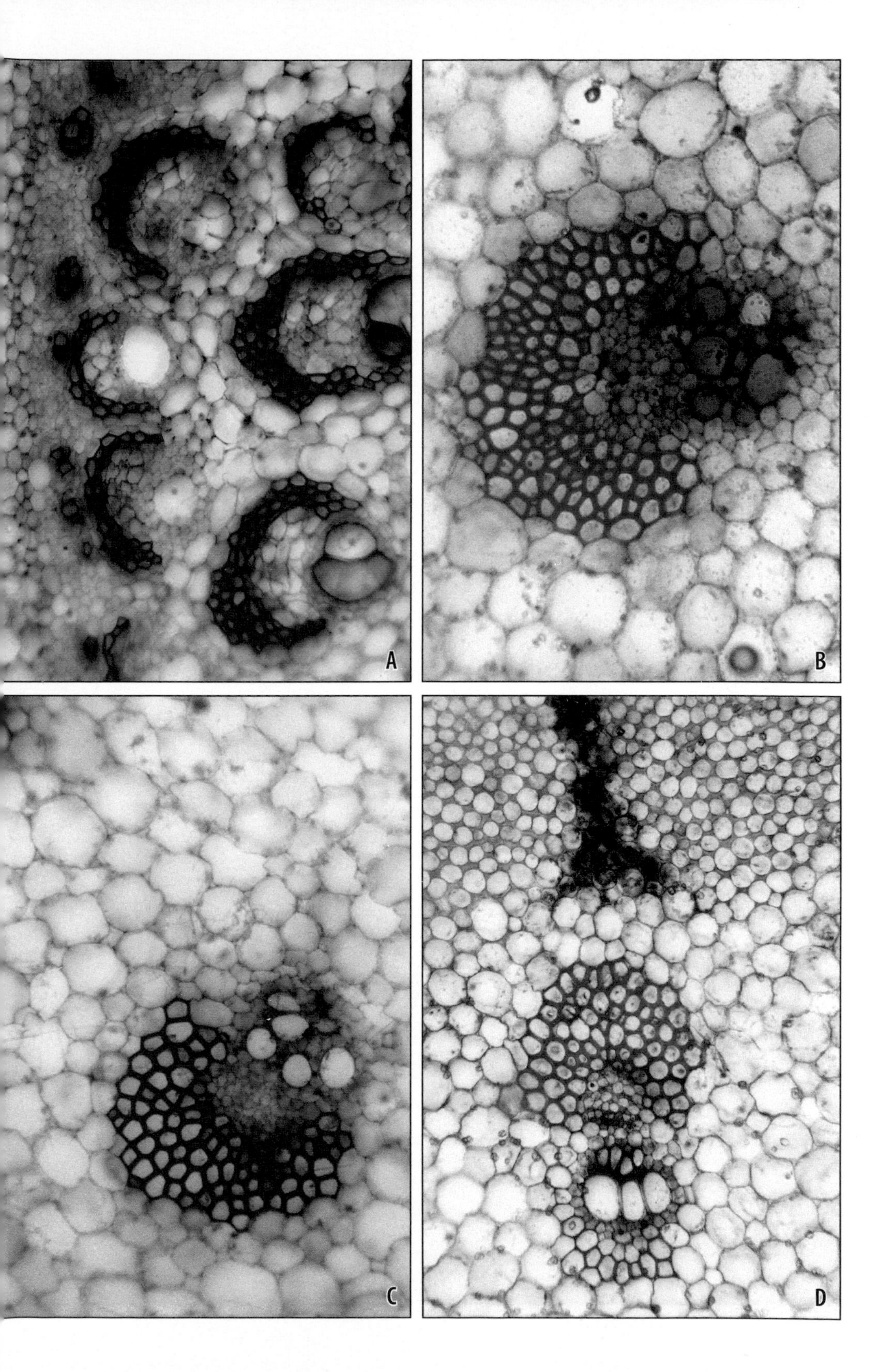
A
B
C
D

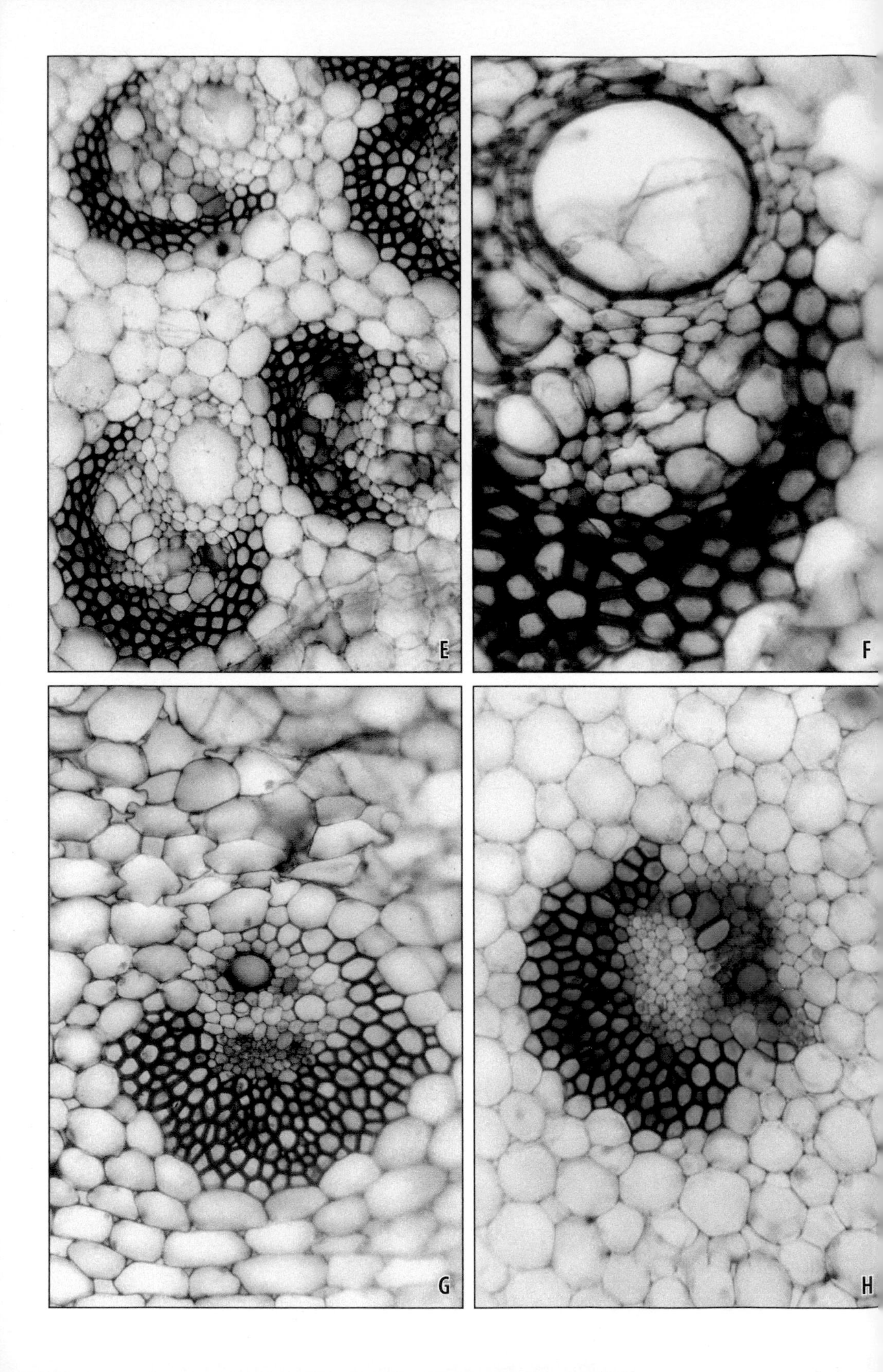
E
F
G
H